The LNICST series publishes ICST's conferences, symposia and workshops.
LNICST reports state-of-the-art results in areas related to the scope of the Institute.
The type of material published includes

- Proceedings (published in time for the respective event)
- Other edited monographs (such as project reports or invited volumes)

LNICST topics span the following areas:

- General Computer Science
- E-Economy
- E-Medicine
- Knowledge Management
- Multimedia
- Operations, Management and Policy
- Social Informatics
- Systems

Sébastien Thomassey · Kim-Phuc Tran ·
Frédéric Vanderhaegen · Tsan-Ming Choi
Editors

Responsible Artificial Intelligence and Data Science

First EAI International Conference, RAIDS 2024
Danang, Vietnam, October 22–24, 2024
Proceedings

 Springer

Editors
Sébastien Thomassey
University of Lille
Lille, France

Frédéric Vanderhaegen
Université Polytechnique Hauts-de-France
Valenciennes, France

Kim-Phuc Tran
University of Lille
Lille, France

Tsan-Ming Choi
University of Liverpool Management School
Liverpool, UK

ISSN 1867-8211   ISSN 1867-822X (electronic)
Lecture Notes of the Institute for Computer Sciences, Social Informatics
and Telecommunications Engineering
ISBN 978-3-032-14054-8   ISBN 978-3-032-14055-5 (eBook)
https://doi.org/10.1007/978-3-032-14055-5

Preface

We are delighted to introduce the proceedings of the 1st European Alliance for Innovation (EAI) International Conference on Responsible Artificial Intelligence and Data Science (EAI RAIDS 2024). Hosted by Dong A University, Vietnam, the conference was held on October 22–24, 2024. This conference brought together researchers, developers, and practitioners around the world who are leveraging and developing state-of-the-art technology in Trustworthy and Responsible AI and Data Science.

The technical program of EAI RAIDS 2024 consisted of 16 accepted papers selected for publication in the proceedings. The conference tracks were: Track 1 - AI for healthcare and medical applications; Track 2 - AI for Human Services and Social Applications; and Track 3 - Theorical contributions for responsible AI. These papers were selected from 39 submissions. Each submission was reviewed following a double-blind process with a minimum of 3 reviews per paper. Aside from the high-quality technical paper presentations, the technical program also featured one panel talk and one poster session. The panel talk provided an overview of current issues and future works on Responsible AI and was composed of three invited guests, Prof. Frederic Vanderhaegen from LAMIH, University of Valenciennes, France, Prof. Xianyi Zeng from GEMTEX, ENSAIT, University of Lille, France, Ass. Prof. Guillaume Tartare from GEMTEX, ENSAIT, University of Lille, France, and the moderator of the session, Prof. Sébastien Thomassey, from GEMTEX, ENSAIT, University of Lille, France. The poster session included PhD works in progress in the field.

Coordination with the local chair, Minh Sam Luong, was essential for the success of the conference. It was also a great pleasure to work with such an excellent organizing committee team for their hard work in organizing and supporting the conference. In particular, the Technical Program Committee members completed the peer-review process of technical papers and made a high-quality technical program. We are also grateful to all the authors who submitted their papers to the RAIDS 2024 conference. Finally, we extend our sincere gratitude to Dong A University, the International Research Institute for Artificial Intelligence and Data Science (IAD), and the International Chair in Data Science and Explainable Artificial Intelligence for their valuable logistical and scientific contributions to the conference.

We strongly believe that the 1st EAI RAIDS conference provided a good forum for all researchers, developers, and practitioners to discuss all science and technology aspects that are relevant to Trustworthy and Responsible AI and Data Science.

Sébastien Thomassey
Kim-Phuc Tran
Frédéric Vanderhaegen
Tsan-Ming Choi

Organization

Steering Committee

Sébastien Thomassey   University of Lille, France
Kim Phuc Tran   University of Lille, France

Organizing Committee

General Chair

Sébastien Thomassey   University of Lille, France

General Co-chair

Kim Phuc Tran   University of Lille, France

TPC Chair and Co-chair

Frédéric Vanderhaegen   Polytechnic University of Hauts-de-France, France
Tsan-Ming Choi (Jason)   University of Liverpool Management School, UK

Sponsorship and Exhibit Chair

Phuong Hanh Tran   Dong A University, Vietnam

Local Chair

Minh Sam Luong   Dong A University, Vietnam

Workshops Chair

Huu Du Nguyen   Hanoi University of Science and Technology, Vietnam

Publicity and Social Media Chair

Marie-Pierre Delespierre   University of Lille, France

Publications Chair

Cédric Heuchenne   University of Liège, Belgium

Web Chair

Tho Nguyen   Dong A University, Vietnam

Posters and PhD Track Chair

Thi Anh Dao Nguyen   Dong A University, Vietnam

Panels Chair

Ludovic Koehl   University of Lille, France

Demos Chair

Guillaume Tartare   University of Lille, France

Tutorials Chair

Quoc Thong Nguyen   Dong A University, Vietnam

Technical Program Committee

Adel Ahmadi Nadi	University of Waterloo, Canada; University of Lille, France
Aamir Saghir	Mirpur University of Science and Technology, Pakistan
Ali Raza	Honda Research Institute EU, Germany
Al-Sakib Khan Pathan	Independent University, Bangladesh
Amine Rghioui	Ibn Zohr University, Morocco
Arne Johannssen	Universität Hamburg, Germany
Ashutosh Singh	Leeds University Business School, UK
Asma Amdouni	Tunis University, Tunisia
Bilal Ahmed Lodhi	Ulster University, UK
Carmen Kar Hang Lee	Singapore University of Social Sciences, Singapore
Cédric Heuchenne	University of Liège, Belgium
Cuong Pham V.	Hanoi University of Public Health, Vietnam
Duc Phong Nguyen	Dong A University, Vietnam
Eric K.H. Leung	University of Liverpool, UK
Fatima Sehar Zaidi	Guangzhou University, China
Florin Popentiu Vladicescu	Politehnica University of Bucharest, Romania
Francisco Falcone	Universidad Pública de Navarra, Spain
Gérard Dray	IMT Mines Alès, France
Giovanni Celano	University of Catania, Italy
Guillaume Tartare	University of Lille, France
Hai Canh Vu	Université de Technologie de Compiègne, France
Hau-Ling Chan	Hong Kong Polytechnic University, China
Helen Ma	Hang Seng University of Hong Kong, China
Huu Du Nguyen	Hanoi University of Science and Technology, Vietnam
Imran Khan	Cardiff University, UK
Inam Ullah Khan	National University of Technology, Pakistan
Jaime Lloret	Universidad Politécnica de Valencia, Spain
Jawad Ahmad	Edinburgh Napier University, UK
Jason J. Jung	Chung-Ang University, South Korea
Jinlian Hu	City University of Hong Kong, China

Kieu-Ha Phung	Hanoi University of Science and Technology, Vietnam
Kim Duc Tran	Dong A University, Vietnam
Kobi Abdessamad	University of Angers, France
Le Minh Thi	Hanoi University of Public Health, Vietnam
Ludovic Koehl	University of Lille, France
Mai Huong Bui	Ho Chi Minh City University of Technology, Vietnam
Michael Boon Chong Khoo	Universiti Sains Malaysia, Malaysia
Moussab Orabi	Rosenberger Group, Germany
Muhammad Atif Ur Rehman	Manchester Metropolitan University, UK
Muhammad Imran	Guangzhou University, China
Muhammad Salah ud din	Hongik University, South Korea
Muhammad Waqar	University of Suffolk, UK
Nhu-Ngoc Dao	Sejong University, South Korea
Nguyen Thanh Duc	McGill University, Canada
Phuong Khanh Thi Nguyen	Tarbes National School of Engineering (ENIT), France
Qinglin Sun	Nankai University, China
Quoc Thong Nguyen	Sofft Industries Groupe Highfi, France; Dong A University, Vietnam
Qurat-Ul-Ain Khaliq	Higher Education Department Lahore, Pakistan
Rakitzis Athanasios	University of Piraeus, Greece; University of the Aegean, Greece
Ramla Saddem	University of Reims Champagne-Ardenne, France
Rehmat Ullah Khan	Cardiff Metropolitan University, UK
Robab Afshari	University of Zanjan, Iran
Safiullah Khan	Manchester Metropolitan University, UK
Sarah Ben Othman	University of Lille, France
Seyedeh Azadeh Fallah Mortezanejad	Ferdowsi University of Mashhad, Iran
Sinh Do	Dong A University, Vietnam
Slim Hammadi	Centrale Lille, France
Surajit Bag	Léonard de Vinci Pôle Universitaire, France
Thang Vu Nguyen	Hanoi University of Science and Technology, Vietnam
Thi Hien Nguyen	CY Cergy Paris Université, France
Thi Thuy Duong Pham	Dong A University, Vietnam
Thu Huong Truong	Hanoi University of Science and Technology, Vietnam
Thuy Thi Nguyen	Vietnam National University of Agriculture, Vietnam
Toan Luu Duc Huynh	Queen Mary University of London, UK

Tri-Hai Nguyen   Van Lang University, Vietnam
Tuyet Nguyen   Ho Chi Minh City University of Technology and
                             Education, Vietnam
Vibhushinie Bentotahewa   Cardiff Metropolitan University, UK
Wei Lin Teoh   Heriot-Watt University Malaysia, Malaysia
Yi-Kuei Lin   National Yang Ming Chiao Tung University,
                             Taiwan
Zhenglei He   South China University of Technology, China

Contents

AI for Human Services and Social Applications

Theorical Contributions for Responsible AI

AI for Healthcare and Medical Applications

Deep Convolutional K-Means of 3D Morphologies of Human Legs for the Implementation of Adaptive Leg Morphotypes and Medical Compression Stockings

Timea Banfalvi[1](✉), Dac Hieu Nguyen[2,3], Kim Duc Tran[1,2],
Guillaume Tartare[1], Pascal Bruniaux[1], Long Hai Le[2,4], Xianyi Zeng[1],
and Kim Phuc Tran[1]

[1] University of Lille, ENSAIT, ULR 2461 - GEMTEX - Génie et Matériaux Textiles,
59000 Lille, France
`timea.banfalvi@ensait.fr`
[2] International Chair in DS and XAI, International Research Institute for Artificial
Intelligence and Data Science, Dong A University, Danang, Vietnam
`ductk@donga.edu.vn`
[3] Institute of Artificial Intelligence and Data Science,, Thuyloi University,
Hanoi, Vietnam
[4] Faculty of Electrical and Electronics Engineering, Dong A University,
Danang, Vietnam

Abstract. This study introduces a novel deep learning framework, the Deep Convolutional K-Means (DCKM), designed to cluster 3D morphologies of human legs for applications in medical compression stockings and adaptive leg morphotypes. By leveraging Convolutional Neural Networks (CNNs) and integrating K-means clustering into the Deep Convolutional Transform Learning (DCTL) framework, the method eliminates the need for a decoder network, reducing overfitting risks and computational complexity. The proposed method efficiently processes 3D point clouds generated from mesh representations of human legs, identifying four distinct morphological clusters. Each cluster is described based on geometric and anthropometric landmarks, enabling more accurate classification of leg shapes. The results highlight the potential of DCKM in medical applications, especially for improving compression stocking designs by adapting to specific leg morphologies. Despite its success, challenges related to transparency and interpretability in unsupervised clustering remain. Future research will focus on enhancing the interpretability of DCKM and expanding the dataset to include pathological leg shapes.

Keywords: 3D morphology · Medical compression stockings · Deep Convolutional K-Mean · leg shape clustering · leg morphotypes

S. Thomassey et al. (Eds.): RAIDS 2024, LNICST 673, pp. 3–14, 2026.
https://doi.org/10.1007/978-3-032-14055-5_1

1   Introduction

The development of deep learning-based methods for various 3D shape representations has been significant since 2015. New deep network architectures have been introduced to process and regenerate 3D shapes for geometric learning [26] In computer vision, 3D data is a resource that contains valuable information about the geometry of an object. [1] There are two main categories of 3D data representation: Euclidean-structured 3D data, which has an underlying grid structure allowing global parametrization, and non-Euclidean 3D data, characterized by an irregular structure, making deep learning a challenging task. According to Ahmed et al. [1], in the Euclidean representation category, enter the descriptors, projections, RGB-D, volumetric, and multi-view, and in the non-Euclidean category, enter the point clouds, grids, and meshes.

In our study, we obtained a database of scanned human subjects, presented at first in point clouds transformed into polygonal meshes that have undergone a pre-treatment process of smoothing, hole-filling, and noise removal. Therefore, they can be considered in their initial form as non-Euclidean data. Instead of translating our 3D shapes into the 2D Euclidean domain (descriptors, projections, multi-view images, etc.), 2.5D domain (RGB D), or 3D Euclidean domain (voxels, octree), we would like to preserve the simplified information that the point could provide, therefor we will opt for deep learning methods that can process this type of data as an input. We choose point clouds over meshes and voxels because, according to the author of [26], point clouds are simple structures that are easier to learn.

Point clouds are spatial representations of 3D objects that proximate their geometry in an unstructured way [2]. Although point clouds do not contain surface information about the 3D object, they still represent the most widely used 3-D data form in reconstruction and classification tasks due to their relatively easy acquisition. PointNet [19] is the first architecture that directly processes point clouds and extracts features for classification tasks by learning spatial encoding of each point, and aggregating individual point features to a global point cloud signature. Qi et al. [20] state that it is important also to know the local structure of points, which PointNet does not exploit. Therefore, they have updated the algorithm to process points "hierarchically sampled in a metric space" and introduced PointNet++. According to Su et al. [23] PointNet, which is not designed to observe local information, cannot describe detailed structural information. They also state that the models mentioned cannot easily describe local regions in 3D point clouds. The authors of [23] propose a deep learning framework for disentangling body type and pose information from 3D point clouds by learning interpretable latent representations of the data in an autoencoder structure, proving that the derived latent features can describe body type. Armstrong et al. [3] proposes an unsupervised approach to partitioning the latent space features of 3D point clouds obtained from a VAE structure and disentangle body type and pose just as in [23]. They describe the intrinsic shape (geometric shape) of point clouds with the Laplace-Beltrami operator and provide a shape descriptor that is invariant to pose changes. Their algorithm can identify the shape of individuals in a different pose without needing labeled data. Point clouds have been studied through multiple deep-learning methods to classify and reconstruct 3D shapes; therefore, in this paper, we will exploit the potential of this form of data in the unsupervised classification of human leg morphologies.

The rest of this paper is organized as follows: We begin by providing an overview of the current literature on human morphological investigations and deep clustering techniques in Sect. 2, where we discuss the importance of understanding 3D morphologies in the context of medical applications, mainly focusing on the design and effectiveness of medical compression stockings. Section 3 introduces the original "Deep Convolutional K-Means," a novel method that integrates K-Means clustering into the Deep Convolutional Transform Learning (DCTL) framework, explaining how it leverages the power of convolutional neural networks (CNNs) to enhance clustering performance. Section 4, titled "Experimental Results," presents the practical application of Deep Convolutional K-Means to our 3D morphology dataset. It includes a detailed explanation of our dataset, the methodology used to calculate the optimal number of clusters, and the performance metrics obtained from the clustering model. The chapter concludes with "Discussion and Conclusion" in Sect. 5, summarizing key insights and future directions in clustering detection for adaptive leg morphotypes and Medical Compression Stockings.

2   Related Work and Background

This paper delves into the existing body of literature concerning human morphological studies aimed at identifying shape variances of the body silhouette. Anthropometry has long been a fundamental tool for scientists to quantify the shape and size of the human body. Although useful, traditional manual measurement tools such as tapes and calipers are time-consuming and less precise. 3D scanners have been introduced to address these limitations, capable of capturing hundreds of anthropometric measures within seconds, with unprecedented precision. Consequently, 3D surface anthropometry has emerged as the most prevalent methodology for studying the geometry and morphology of external human body tissues [13]. Numerous contemporary studies employ 3D surface anthropometry to quantify human shapes. One of the earliest 3D anthropometric surveys, known as the CAESAR study (Civilian American and European Surface Anthropometry Resource) [21], was conducted in the United States and Europe, encompassing scans of over 5000 individuals. This dataset has been extensively utilized in scientific research to explore the variance in human body shapes. For instance, Azouz et al. [4] analyzed 3D human body shapes using Principal Component Analysis (PCA) and identified five principal components that exhibited the most significant variance within their population of 300 male subjects from the CAESAR dataset. Their research indicated that weight, accounting for 33.86% of the total variation, was the most variable component, followed by posture at 15.11%. Other factors such as muscularity, arm-torso spacing, and head position were also identified as significant variability factors. These results were interpreted as a quantitative generalization of Sheldon's morphotypes (mesomorph, endomorph, ectomorph) [22].

Further studies by Xi et al. [25] demonstrated that segmenting body parts and performing PCA on each segment separately yielded higher variation than analyzing the entire body. For example, PCA on the leg revealed that the first principal component described leg height variation. In contrast, subsequent components described leg straightness, pose along the left-right direction, bending extent, and scale. Multiple sources have employed PCA for shape analysis and the creation of parametric models. Li et al. [14] presented an approach for classifying 3D female torso shapes by identifying five principal components that describe torso shape variation. Additionally, Hamad et al. [12] extracted geodesic shape descriptors from 500 3D female torsos and distinguished three morphotypes through a two-level clustering procedure (SOM followed by k-means), which facilitated the proposal of a new sizing system for French apparel companies. Liu et al. [15] developed a methodology for categorizing lower body shapes and sizes using three shape-describing parameters, presenting three distinct lower limb shapes to design better medical compression stockings for the Chinese population.

In parallel, the field of deep clustering has rapidly evolved within unsupervised learning, attracting significant attention for its potential to manage high-dimensional and complex data effectively. K-means [16] and Gaussian Mixture Models (GMM) [5] are traditional clustering algorithms that are effective for low-dimensional and smaller datasets; however, they encounter difficulties when dealing with high-dimensional and large-scale data due to their sensitivity to noise and dependence on linear transformations. Deep learning methods have been integrated into clustering to overcome these limitations, leading to the development of deep clustering techniques. These methods leverage the powerful feature extraction capabilities of deep neural networks (DNNs), enabling the extraction of highly representative features that facilitate more accurate clustering results [24]. A fundamental method in deep clustering is Deep Embedded Clustering (DEC) [27], which starts with pre-training an autoencoder to initialize the network and subsequently fine-tunes the encoder using clustering-specific loss functions such as the Kullback-Leibler (KL) divergence to enhance clustering performance. Improved Deep Embedded Clustering (IDEC), proposed by Xifeng Guo et al. [8], enhances DEC by incorporating a reconstruction loss to preserve the local structure of the data-generating distribution. Due to its impressive performance in image processing, the convolutional neural network (CNN) has also been employed in deep clustering. Dizaji et al. [6] proposed a method that employs joint convolutional autoencoder embedding and relative entropy minimization to cluster high-dimensional data effectively. This method defines a clustering objective function based on relative entropy minimization, thereby demonstrating the effective use of CNNs. The Deep Convolutional K-Means (DCKM) clustering method [7] improves traditional CNN-based clustering by embedding K-means clustering into a Deep Convolutional Transform Learning (DCTL) framework. This approach avoids overfitting by eliminating the need for a decoder network and achieves improved clustering performance, particularly in scenarios with numerous clusters and limited data samples. Recently, Pooja Gupta et al. [10] introduced an innovative unsupervised multi-channel fusion clustering model named DeConFCluster, based on DCTL. This framework integrates DeConFuse [11] and K-Means clustering modules, training them end-to-end, therefore eliminating the need to learn the weights of decoder or deconvolutional layers and reducing computational overhead. This design helps prevent overfitting, especially in scenarios with limited data instances and many classes.

3   Deep Convolutional K-Means

Deep Convolutional K-Means (DCKM) is a novel clustering method proposed by Anurag et al. [7]. DCKM integrates the capabilities of convolutional neural networks (CNNs) with K-means clustering to create a robust framework for unsupervised learning suited for image data. Traditional CNN-based clustering approaches often rely on encoder-decoder architectures, which may result in overfitting in data-limited situations due to the need for additional weights in the decoder network. DCKM addresses this limitation by introducing a Deep Convolutional Transform Learning (DCTL) that eliminates the decoder, thereby learning fewer parameters and reducing over-fitting potential. The schematic diagram of DCKM is shown in Fig. 1.

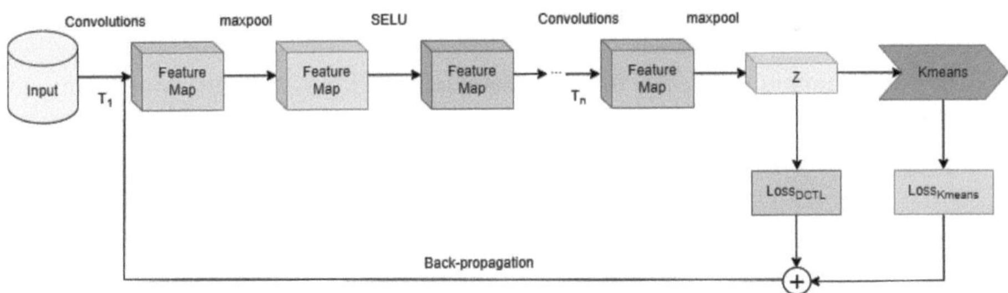

Fig. 1. Schematic diagram of Deep Convolutional K-Means

3.1   Deep Convolutional Transform Learning

Deep Convolutional Transform Learning is an extension of Convolutional Transform Learning (CTL) [17], which learns a set of convolutional filters that produce corresponding representations when applied to the input data. The mathematical formulation of DCTL to solve an optimization problem that aims to learn the filters and representations simultaneously. Optimizing $Loss_{DCTL}$ is expressed as follows

$$\min_{T_1,T_2,T_3,Z} \|T_3*(T_2*(T_1*X))-Z\|_F^2 + \psi(Z) + \lambda \left(\sum_{i=1}^{3}(\|T_i\|_F^2 - \log\det(T_i)) \right) \quad (1)$$

where T_1, T_2, and T_3 are the convolutional filters at different layers, X is the input data, and Z is the representation. The penalty terms $\|T_i\|_F^2$ and $\log\det(T_i))$ prevent trivial and degenerate solutions, ensuring the uniqueness of the learned filters.

3.2 Clustering DCTL with K-Means

Incorporating K-means clustering into the DCTL framework involves embedding the clustering loss directly into the transform learning process, allowing the feature learning and clustering processes to be optimized jointly, resulting in more coherent and well-separated clusters. Initially, images are processed through convolutional filters T_1 followed by max-pooling and activation using the Scaled Exponential Linear Unit (SELU) function. Then, the output result is processed by a second set of convolutional filters T_2 and passed to another round of max-pooling. This process continues until T_n before being input into the K-means clustering algorithm. The expression is shown as:

$$\min_{T_1,T_2,T_3,Z} Loss_{DCTL} + \mu\|Z - ZH^T(HH^T)^{-1}H\|_F^2 \quad (2)$$

Here, the matrix H represents binary indicator variables h_{ij}, where $h_{ij}=1$ if $x_j \in$ cluster i and 0 otherwise.

The combination of DCTL and K-means clustering can significantly improve clustering performance, especially in situations when there are several clusters and a limited amount of samples.

4 Experimental Results

The results of the experimental application of Deep Convolutional K-Means (DCKM) clustering to the 3D morphologies of human legs are presented in this section. We collected a dataset consisting of 127 3D scans of human legs. We collected a dataset consisting of 127 3D scans of human legs, 71 of which are women (precisely 61 under 30 and 10 above the age of 30) and 56 men (49 under 30 and 7 above the age of 30). Each 3D scan underwent a pre-treatment process, which allowed for the correction of eventual geometric anomalies and the removal of noise data. After these processes, we extracted the left leg of each subject with precise segmentation in a CAD environment using anthropometric landmarks, thus obtaining a uniform representation of the leg throughout the dataset.

We used Principal Component Analysis (PCA) to lower the high-dimensional mesh data to a lower-dimensional representation, then the Elbow approach on

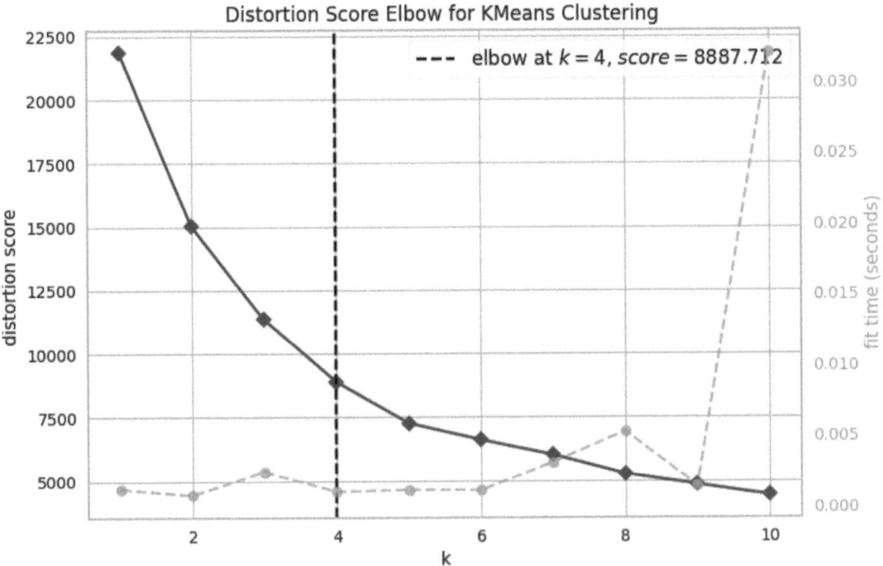

Fig. 2. Using the Elbow method to figure out the optimal number of clusters

the PCA outputs to determine the appropriate number of clusters (K). By plotting the sum of squared errors (SSE) versus the number of clusters (k), the Elbow approach [18] finds the ideal number of clusters, as the SSE decreases as the value of k increases. The experiment indicated that our dataset's optimal number of clusters was four (k = 4). Figure 2 illustrates the Elbow plot used to determine the optimal K.

With the optimal k clusters determined (k = 4), we trained the Deep Convolutional K-Means (DCKM) model using our dataset. Due to meshes being complicated structures made up of vertices, edges, and faces, we converted meshes to 3D point clouds [9], which is necessary for simplified representation, compatibility with deep learning algorithms, specifically 3D CNN, as illustrated in Fig. 3. The DCKM model was configured with the following parameters: $\lambda = 0.001$, $\mu = 1.0$, epochs are 2000, layers are 2, and each layer consists of 8 filters. Our architecture is coded in Python 3.11 and relies on the Pytorch Framework, employing in hardware machine with a chipset Intel i5-13500 CPU @ 3.50GHz, 32GB RAM, 512 SDD, equipped with Nvidia RTX 3060 12GB, and operates on the Windows Subsystem for Linux (WSL) platform. After training the DCKM model, we extracted the resulting features and applied the K-means clustering algorithm with 4 centroids. We archived four distinct clusters, as shown in Fig. 4.

After identifying the 4 clusters, which represent a population of legs with similar shapes, to help us describe the shape of each cluster, we identified the morphotype, which is the subject closest to the group's centroid. We also plotted the circumferential values of 17 cross sections equally distributed on the length of the leg at a normalized height, resulting in a set of curve plots describing the leg's

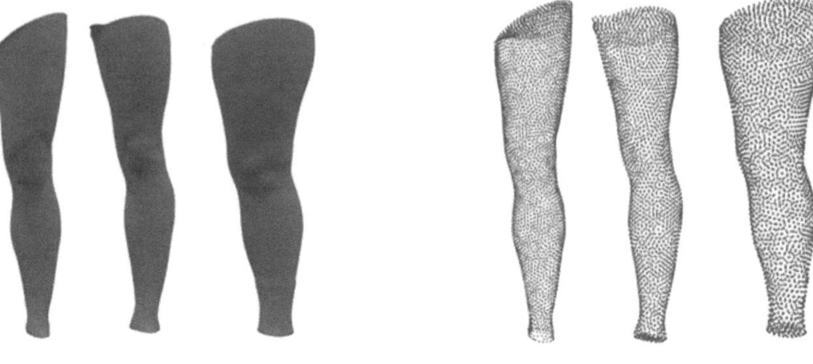

(a) 3D scans of human legs   (b) 3D point clouds of human legs

Fig. 3. Converting 3D scans to 3D point clouds of human legs

volume and some shape characteristics. Figure 5 illustrates the population of the four clusters with the mentioned curve plots, where the red curve represents the morphotype, which is also visible on the left side of each subfigure. The knee region of the shapes in cluster 1 visible in Fig. 5a is padded from both anterior and posterior sides, having an observable bump behind the knee, called the popliteal fat pad. The tight shape is a trapezoid with a lack of muscle development. The lower leg region shows the same underdevelopment in muscle mass but the ankles tend to grow, especially in larger sizes. The tight region of cluster 2, visible in Fig. 5b, is muscly, especially the quadriceps group. From both the anterior and medial view we can observe the head of the vastus medialis that creates a pronounced indentation at the medial side of the knee. The lateral contour of the tight is rounded, and the vastus lateralis is noticeably well-developed, making the contour of the iliotibial tract visible. The tight also rounds towards the front, a sign of a well-developed rectus femoralis. The main difference between males and females is the possible presence of the outer tight fat pad, which is more likely to appear in the case of women. The lower leg is also muscly, as the subjects of this group have larger calf muscles. Cluster 3 on Fig. 5c are in general trapezoid shapes that lack curves; the knee region is padded, especially the under knee region. The inner and outer tights are straight, suggesting a uniform fat distribution. This shape of the leg tends to be underdeveloped in muscle mass and tends to have an even distribution of fat pads along the length of the leg. The shape of cluster 4, visible in Fig. 5d is characterized by a thinner tight and lower leg region, with a very bony knee, and compared to the other clusters, this population shows the thinnest ankle region. For these reasons, we would characterize this shape as lacking in general fat deposits.

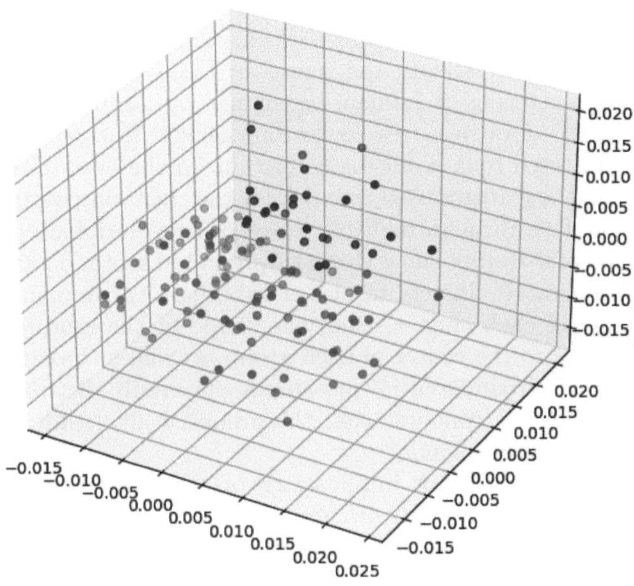

Fig. 4. Cluster visualization in 3D

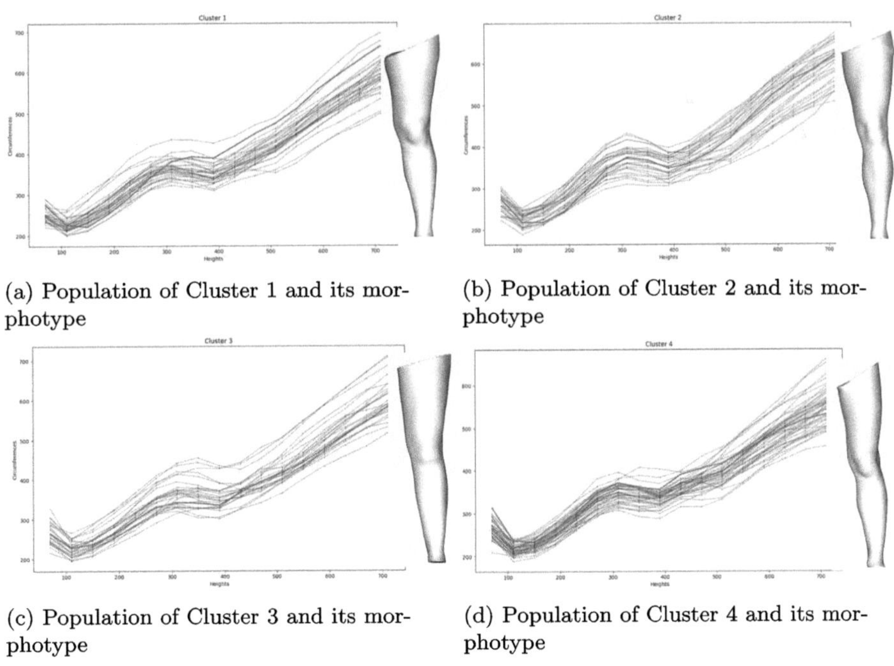

(a) Population of Cluster 1 and its morphotype

(b) Population of Cluster 2 and its morphotype

(c) Population of Cluster 3 and its morphotype

(d) Population of Cluster 4 and its morphotype

Fig. 5. Extracted morphotypes and cluster population described with the interpolation between heights and circumferences of 17 cross sections on the leg

5   Discussion and Conclusion

This study aimed to investigate the power of Deep Convolutional Transform Learning in the clustering of human leg shapes by finding an optimal representation of the 3D data in a new lower-dimensional feature space, which can ultimately be clustered by a K-means algorithm. In the Deep Convolutional K-means algorithm, the features used as input for the clustering were automatically learned from the data through multiple layers of convolutional operations.

Our experimental results show that the presented method can distinguish 4 relatively well-separated clusters from a database of 127 human legs. By evaluating the shape of each cluster using the shape of the morphotype and the five closest subjects to the center of each group, as well as looking at the curve plots defined by 17 circumferential measurements evenly distributed along the length of the leg, we can conclude that the proposed algorithm can distinguish leg shapes using 3D point clouds.

One of the most significant implications of these findings is the potential to revolutionize the field of medical compression stockings by providing shape categories that can be used in the proposition of a new sizing system for the industry. Another valuable use of these results lies in the potential to aid the design of these products by taking the mean shape of each determined group and modeling the medical product directly on this shape, creating a more reliable and better-fitting device, aiding the treatment of Chronic Venous Insufficiency and Lymphoedema.

However, this study is not without its limitations. While the proposed deep learning method is promising results in finding complex patterns, it also has to be mentioned that there is a lack of information on how and why those data points got grouped, causing the well-known "black-box" effect. This lack of transparency makes it difficult to describe the four shape categories defined by the algorithm directly, therefore in our future studies, we will be working on the interpretability of the proposed method.

Another call to identifying lower-dimensional features from 3D point clouds automatically hand; even through data augmentation, the results can not be generalizable. To solve this problem, we will continue to enrich our database with healthy legs and legs with the pathology present, including the geometry of lymphoedema patients in our research.

In conclusion, this study has demonstrated the potential of Deep Convolutional K-means to automatically identify lower-dimensional features from 3D point clouds and successfully cluster human leg shapes in 4 distinct clusters. These findings hold significant promise for the medical compression stocking industry, offering a foundation for developing a new sizing system and improving product design to heal venous disorders.

References

1. Ahmed, E., et al.: A survey on deep learning advances on different 3D data representations. arxiv (2018). arXiv preprint arXiv:1808.01462 (2019)
2. Ahmed, E., et al.: A survey on deep learning advances on different 3D data representations. arXiv (2019). http://arxiv.org/abs/1808.01462 Accessed 16 Feb 2024
3. Aumentado-Armstrong, T., Tsogkas, S., Jepson, A., Dickinson, S.: Geometric disentanglement for generative latent shape models. arXiv (2019). http://arxiv.org/abs/1908.06386 Accessed 13 Sep 2024
4. Azouz, Z.B., Rioux, M., Shu, C., Lepage, R.: Characterizing human shape variation using 3D anthropometric data. Visual Comput. **22**(5), 302–314 (2006). https://doi.org/10.1007/s00371-006-0006-6 see reference [5]
5. Pattern recognition and machine learning. ISS. Springer, New York (2006). https://doi.org/10.1007/978-0-387-45528-0_9
6. Dizaji, K.G., Herandi, A., Deng, C., Cai, W., Huang, H.: Deep clustering via joint convolutional autoencoder embedding and relative entropy minimization. In: 2017 IEEE International Conference on Computer Vision (ICCV), pp. 5747–5756 (2017). https://doi.org/10.1109/ICCV.2017.612
7. Goel, A., Majumdar, A., Chouzenoux, E., Chierchia, G.: Deep convolutional k-means clustering. In: 2022 IEEE International Conference on Image Processing (ICIP), pp. 211–215 (2022). https://doi.org/10.1109/ICIP46576.2022.9897742
8. Guo, X., Gao, L., Liu, X., Yin, J.: Improved deep embedded clustering with local structure preservation. In: Proceedings of the Twenty-Sixth International Joint Conference on Artificial Intelligence. IJCAI-2017, International Joint Conferences on Artificial Intelligence Organization (2017). https://doi.org/10.24963/ijcai.2017/243
9. Guo, Y.: Deep learning for 3D point clouds: a survey. IEEE Trans. Pattern Anal. Mach. Intell. **43**(12), 4338–4364 (2021). https://doi.org/10.1109/TPAMI.2020.3005434
10. Gupta, P., Goel, A., Majumdar, A., Chouzenoux, E., Chierchia, G.: Deconfcluster: deep convolutional transform learning based multiview clustering fusion framework. Sig. Process. **224**, 109597 (2024). https://doi.org/10.1016/j.sigpro.2024.109597, https://www.sciencedirect.com/science/article/pii/S0165168424002160
11. Gupta, P., Maggu, J., Majumdar, A., Chouzenoux, E., Chierchia, G.: DeConFuse: a deep convolutional transform-based unsupervised fusion framework. EURASIP J. Adv. Sig. Process. **2020**(1), 1–32 (2020). https://doi.org/10.1186/s13634-020-00684-5
12. Hamad, M., Thomassey, S., Bruniaux, P.: A new sizing system based on 3D shape descriptor for morphology clustering. Comput. Ind. Eng. **113**, 683–692 (2017). https://doi.org/10.1016/j.cie.2017.05.030, see reference [8]
13. Jones, P.R.M., Rioux, M.: Three-dimensional surface anthropometry: applications to the human body. Opt. Lasers Eng. **28**(2), 89–117 (1997). https://doi.org/10.1016/S0143-8166(97)00006-7, see reference [3]
14. Li, P., Mitchell, K.B.: A shape classification scheme for female torso. Appl. Ergon. **106**, 103904 (2023). https://doi.org/10.1016/j.apergo.2022.103904, see reference [7]
15. Liu, R., et al.: Stratified body shape-driven sizing system via three-dimensional digital anthropometry for compression textiles of lower extremities. Text. Res. J. **88**(18), 2055–2075 (2018). https://doi.org/10.1177/0040517517715094, see reference [9]

16. MacQueen, J.: Some methods for classification and analysis of multivariate observations. In: Proceedings of the 5th Berkeley Symposium on Mathematical Statistics and Probability, pp. 281–297. Statistics, University of California Press, Berkeley, Cham (1967)
17. Maggu, J., Chouzenoux, E., Chierchia, G., Majumdar, A.: Convolutional transform learning. In: Cheng, L., Leung, A.C.S., Ozawa, S. (eds.) Neural Information Processing, pp. 162–174. Springer International Publishing, Cham (2018)
18. Marutho, D., Handaka, H.S., Wijaya, E., Muljono: The determination of cluster number at k-mean using elbow method and purity evaluation on headline news. In: 2018 International Seminar on Application for Technology of Information and Communication, pp. 533–538 (2018). https://doi.org/10.1109/ISEMANTIC.2018. 8549751
19. Qi, C.R., Su, H., Mo, K., Guibas, L.J.: Pointnet: deep learning on point sets for 3D classification and segmentation. arXiv (2017). http://arxiv.org/abs/1612.00593 Accessed 13 Sep 2024
20. Qi, C.R., Yi, L., Su, H., Guibas, L.J.: Pointnet++: deep hierarchical feature learning on point sets in a metric space. arXiv (2017). http://arxiv.org/abs/1706.02413 Accessed 13 Sep 2024
21. Robinette, K.M., Daanen, H., Paquet, E.: The CAESAR project: a 3-D surface anthropometry survey. In: Second International Conference on 3-D Digital Imaging and Modeling (Cat. No.PR00062), pp. 380–386. IEEE Comput. Soc, Ottawa, Ont., Canada (1999). https://doi.org/10.1109/IM.1999.805368, see reference [4]
22. Sheldon, W.H.: The varieties of human physique: an introduction to constitutional psychology. Harper, New York (1940). http://archive.org/details/ varietiesofhuman0000shel Accessed 10 July 2024. see reference [2]
23. Su, F.G., Lin, C.S., Wang, Y.C.F.: Learning interpretable representation for 3D point clouds. In: 2020 25th International Conference on Pattern Recognition (ICPR), pp. 7470–7477 (2021). https://doi.org/10.1109/ICPR48806.2021.9412440
24. Wei, X., Zhang, Z., Huang, H., Zhou, Y.: An overview on deep clustering. Neurocomputing **590**, 127761 (2024) https://doi.org/10.1016/j.neucom.2024.127761, https://www.sciencedirect.com/science/article/pii/S0925231224005320
25. Xi, P., Lee, W.S., Shu, C.: A data-driven approach to human-body cloning using a segmented body database. In: 15th Pacific Conference on Computer Graphics and Applications (PG'07), pp. 139–147. IEEE, Maui, HI, USA (2007). https://doi.org/ 10.1109/PG.2007.45, see reference [6]
26. Xiao, Y.P., Lai, Y.K., Zhang, F.L., Li, C., Gao, L.: A survey on deep geometry learning: from a representation perspective. Comput. Visual Media **6**(2), 113–133 (2020). https://doi.org/10.1007/s41095-020-0174-8
27. Xie, J., Girshick, R., Farhadi, A.: Unsupervised deep embedding for clustering analysis. In: Balcan, M.F., Weinberger, K.Q. (eds.) Proceedings of The 33rd International Conference on Machine Learning. Proceedings of Machine Learning Research, vol. 48, pp. 478–487. PMLR, New York, New York, USA (20–22 Jun 2016). https://proceedings.mlr.press/v48/xieb16.html

Unsupervised Classification of 3D Morphologies of Human Legs for the Implementation of Adaptive Leg Morphotypes and Medical Compression Stockings: A Survey and Perspective

Timea Banfalvi[1(✉)], Kim Duc Tran[1,2], Guillaume Tartare[1], Pascal Bruniaux[1], and Kim Phuc Tran[1]

[1] University of Lille, ENSAIT, ULR 2461 - GEMTEX - Génie et Matériaux Textiles, 59000 Lille, France
timea.banfalvi@ensait.fr
[2] International Chair in DS and XAI, International Research Institute for Artificial Intelligence and Data Science, Dong A University, Danang, Vietnam
ductk@donga.edu.vn

Abstract. Shape and size are factors that influence the way a medical compression stocking exerts pressure on the leg. To create a correct medical device, not only the mechanical properties of the textile structure are relevant but also the way the knitting pattern is being created, the latter being influenced by the morphological shape of the wearer. In this study, we present a comprehensive shape analysis framework for a dataset of human legs, focusing on quantifying the shape variations within our population, leveraging both classical and modern computational techniques. Initially, we process 1D and 2D shape descriptors with spectral clustering methods to identify morphological groups for human legs. This work also presents a methodology for extracting a 3D shape descriptor with promising comparability factors alone. Our future work aims to take the 3D leg as a polygonal mesh, introduce it in known mesh convolutional neural networks, and determine a final set of morphotypes. This could enhance our understanding of leg shape and aid in the design of well-adapted medical compression stockings.

Keywords: Human leg morphology · unsupervised classification · shape descriptors · medical compression stockings · Spectral clustering

1 Introduction

Compression therapy is used worldwide to improve blood circulation and reduce tissue inflammation and lymphatic fluid accumulation [27]. To treat Chronic Venous Insufficiency (CVI) and Lymphoedema an optimal compression level has to be determined [23], which is one of the biggest challenges for both medical

S. Thomassey et al. (Eds.): RAIDS 2024, LNICST 673, pp. 15–34, 2026.
https://doi.org/10.1007/978-3-032-14055-5_2

professionals and industrial parties. The variation of the shape and size of the leg is highly influenced by these two diseases [6,17] but also in general legs have very different shapes, which makes it hard for the industry to design Medical Compression Stockings that correspond to the real leg shape and also suit everyone. Instead, they consider the leg in theoretical modeling as a conical shape that has nothing to do with the real anatomy of the wearer.

This project aims to revolutionize the design and manufacturing of medical compression stockings (MCS's) by a series of innovative objectives. First of all by creating an extensive database with a 3D scanning technology, that can capture the surface geometry of the subjects. This technology also allows us to capture important anthropometric points and measurements focusing on the lower limb. By analyzing these landmarks and measurements we create a shape descriptor that can capture the leg geometry in a simplified manner. This shape descriptor can be further studied with unsupervised classification methods to identify morphological groups, from which we aim to identify the morphotype, a shape that describes the population, to which the leg shape of every member of the morphological group should resemble. Additionally, we will create an adaptive morphotype for each identified group, that would adjust in size (volume and height) to a predefined sizing table. This adaptive morphotype can be used to create an MCS model with an ease model approach, generating a surface that is precisely adapted to the leg's shape, this way we can not only parameterize the geometry of the wearer but also have an influence on the pressure that needs to be applied to that specific shape. The dimensional parameters derived from this digital parametric model will guide the manufacturing process for better MCS design.

In this paper, we present our initial results derived from the unsupervised classification of a 1D (measurements) and a 2D (interpolation graphs) shape descriptor. These results have granted us valuable information about shape and guided us toward the next stage of our research, which is the extraction of a 3D shape descriptor. The rest of this paper is structured as follows: Sect. 2 provides a detailed review of related work and existing methodologies in 3D geometric learning. In Sect. 3, we present our first results obtained with the unsupervised classification of a 1D and a 2D shape descriptor of the lower limb. Section 4 presents a new approach to extracting a 3D shape descriptor. In Sect. 5, we present the difficulties and challenges encountered in our research as well as the perspectives that emerge from our findings. Finally, Sect. 6 concludes the paper by summarizing our findings and proposing directions for future research.

2   State of the Art

Compression therapy, is one of the oldest treatment options [22] in the management of chronic venous insufficiency and lymphoedema and it happens most often in the form of medical compression stockings [30,32]. Chronic venous insufficiency is a disease that is touching a large population worldwide [28], the risk of its appearance increases with age and it is more common in the case of women.

Sitting and standing professions also contribute to the possibility of CVI [26, 29] and for preventive treatment medical professionals suggest the use of light compression stockings, which could improve venous and lymphatic flow, if the compression stockings are optimally fitted. Schupke et al. [29], mentions that in clinical practice the optimal fitting of compression stockings is not always correct. Body size and body shape information are both necessary to know the accurate geometry of a target human body [9], but to fit medical compression stockings, the only taken measurements are the circumferences and heights of the ankle, calf, and upper tight [2], that indeed tells the size of the leg but does not tell much about the shape of the leg. Having a poorly fitted medical device can on one hand exert the wrong pressure on the leg and on the other hand it can cause pain and discomfort for the wearer, resulting in non-compliance. Medical compression stockings are tight-fitted on the leg, therefore the shape of this body part influences greatly the applied interface pressure. Both Venous Insufficiency and Lymphoedema are two chronic diseases, which can cause severe deformation of the leg's shape, and depending on the type and severity of these diseases, the contact pressure needs to be controlled effectively. Many factors can influence this, one of which is proven to be the morphology of the body. Ilska et al. [13] present the importance of knowing the body circumferences with and without MCS, to correctly estimate the pressure applied on the soft body tissue because surface pressure leads to some changes in the circumferences and their lengths, the author also concluded that for a correct design of the compression products it is highly important to take into consideration the impact of body on the value of unit pressure to match the intended and actual pressure value exerted by the product. The concept of personalization of compression products has been proposed in [13], suggesting that compression therapy is the most effective when individual anthropometric measurements are considered, [19] recommends personalization of the MCS for patients with specific morphological features. Xi et al. [34] proposes a parametric model, that characterizes the shape of human calf, also in the concept of personalization of MCS. For the correct treatment of CVI and Lymphoedema, patients have to wear the medical device for prolonged periods, and studies [4, 7, 16, 21, 28, 33], show that the compliance rate of medical compression stocking is relatively low for multiple reasons. Franks et al. [9] reveal that some patients are suffering from skin irritation after just one week of wearing MCS, 15% of the studied patients could not put their stockings on at all, and 26% of them were also having great difficulty with donning the products. The great discomfort, in the case of a minority of the patients, comes from itchiness, swelling, and rashes caused by the MCS. Erickson et al. [8] also reveals the same causes for non-compliance and the author concludes that strict compliance to treatment protocol is essential to decrease the time of healing venous ulcers. According to Kankam et al., [16] non-adherence is due to pain, discomfort, difficulty donning stockings, tightness, feeling hot, perceived ineffectiveness, skin irritation, cost, and lack of cosmetic appeal. Not being independent in applying the product is one of the main reasons why patients refuse or avoid to wear compression stockings [4], even though there exist various tools on the market that could help the donning of the product. In the study of Jindal et al. [14], they point out that commercialized MCS brands, designed and manufactured in Europe should adapt their product sizing to the morphology of the target country, where they sell the product, the study proved that within the three available MCS brands in India, none of them were truly correct for the morphology of Indian population, causing problems of donning and slippage of the product. The research of Liu et al. [19] was also conducted for the same reason, in the case of the Chinese population, where the commercialized brands were not adapted to the country's body morphology, therefore the author proposed a new sizing system, identifying lower limb morphologies to offer the industry a basis for product design. According to Nørregaard [25], health practitioners have great difficulty prescribing off-the-shelf products for patients suffering from leg ulcers only with the three-point measuring method (instep, ankle, calf). For these reasons, the reevaluation of the size of medical compression stockings, also taking into consideration the anatomical shape, is crucial to provide better fitting and more precise medical devices for patients suffering from these chronic diseases.

3 Unsupervised Classification of Morphologies of Human Legs: First Results

Data Acquisition 1st Measurement Campaign. To improve our previous results presented in [3], we have enlarged our previous set of 127 subjects to a total of 251 subjects, using 3D scanning technology. 3D surface anthropometry has been used to collect anthropometric measurements and visualize body shape. The 251 individuals in our database are composed of 152 women and 99 men and these 3D scans have been pre-treated with a process that allowed us to correct the surface anomalies. We have collected 17 circumferential measurements equally distributed between the ankle and the crotch and their respective heights.

We created a curve representation of the the 17 circumferential values and their heights, and we obtained the graphic in Fig. 1 (a). In this figure, we can visualize the first form of the 2D shape descriptor of two subjects with different heights taken from our database. It can be observed that the corresponding data points do not match the height between the two individuals. This height difference between the two persons makes it impossible to compare the two shapes accurately. To resolve this mismatch, we implemented a normalization technique, using the minimal bounding box to overlap all the shapes in a comparable manner, which can be seen in Fig. 1 (b). Figure 1 (c) shows the final set of descriptors for the entire database.

Principal Component Analysis (PCA). With the purpose of reducing the complexity of our data and to insure 3D visualisation we have used Principal Component analysis. Our data is composed of 251 observations and 18 variables. To obtain the 3D subspace of the 18D data, as a first step we have identified the 5 variables that contained the most variation (ankle at lateral malleolus (V1);

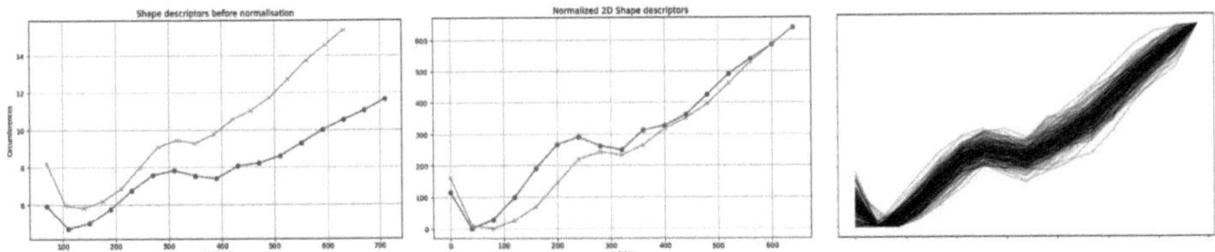

Fig. 1. Construction of the 2D shape descriptor; an example of two subjects with different heights, (a) represented with the original leg length and leg circumferences, (b) with minimal binding box normalization, and (c) the overall database of normalized 2D shape descriptors

ankle at it's thinnest point (V2); mid lower leg (V4); calf (V7); tight (V16)) in the dataset and we have eliminated the rest of 13 redundant variables. The linear combination of these variables resulted in 3 Principal Components that can describe 97.4% of the total variations. WE have made the same observation as Jolliffe in [15], that the first PC describes the overall 'size' difference between the circumferences, the second PC describes the ankle variation and similarly the third PC describes the tight variation between the subjects.

Clustering. To obtain the morphological groups we applied Spectral Clustering algorithm to group together similar leg shapes. Luxburg describes the basic steps of spectral clustering in [24], and as she describes the first step is the computation of the similarity matrix. The weighted adjacency matrix has been computed and plotted in for of a graph, based on the Euclidean distance between the data points. We found the ideal number of neighbors by trying out multiple values of k and visualizing the closeness and separation of data points, and the results proved to be the best in case of 3 neighbors for the 1D measurements and 4 neighbors for the 2D curve plots. With the help of the correct weighted adjacency matrix, we computed the normalized random walk Laplacian matrix, extracted the first k (k = nr.neighbors) eigenvectors that are assigned to the smallest eigenvalues and we applied K-means on these values, which gave us the ultimate clusters.

To verify the quality of the spectral clustering algorithm and at the same time to determine the optimal number of clusters we used two methods. First, we used Rousseeuw's method [31], called the Silhouette score, which is a graphical representation of the predicted clusters which measures the tightness and separation of the detected clusters. By computing the overall silhouette scores for a range of k, we can observe a peak point at k=4 for both types of classifications, which can be visualized in Figs. 2a and 2c. We have tested a second method, called the Davies-Bouldin Index, that we have also computed for a range of k, this time observing a low point in k=4 for both cases. These two methods both suggest that four clusters can be detected in our data no matter of the descriptor's type.

Figure 3a and Fig. 3b allowed us to visualize the four determined clusters and the 251 data points in the 3D spectral embedding for both of the shape

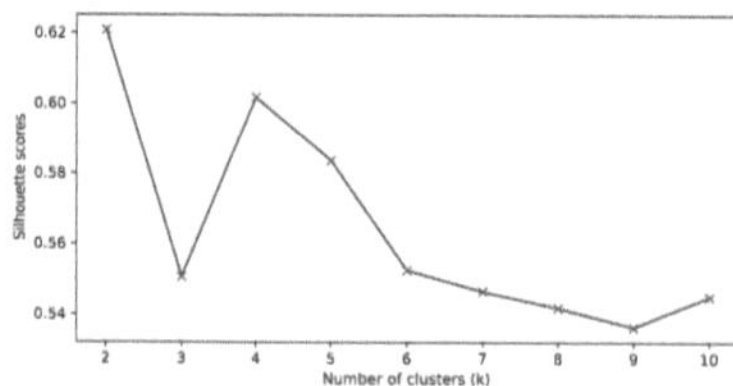

(a) Overall Silhouette average for the 1D shape descriptor clustering

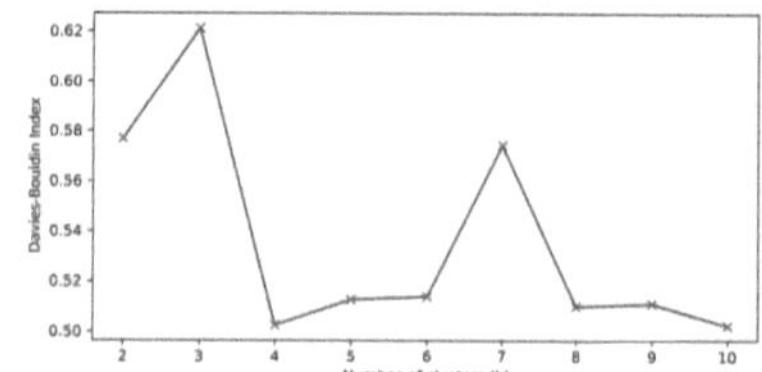

(b) Davies-Bouldin Index of the 1D shape descriptor

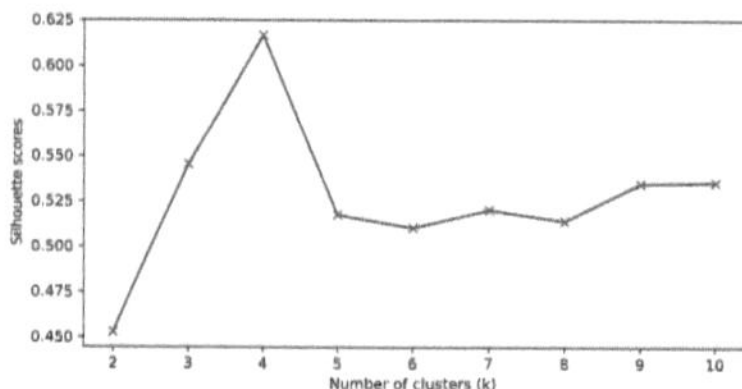

(c) Overall Silhouette average for the 2D shape descriptor clustering

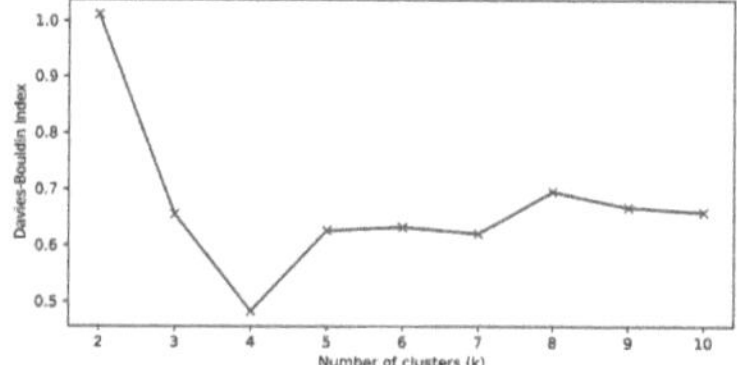

(d) Davies-Bouldin Index of the 2D shape descriptor

Fig. 2. Cluster quality indexes

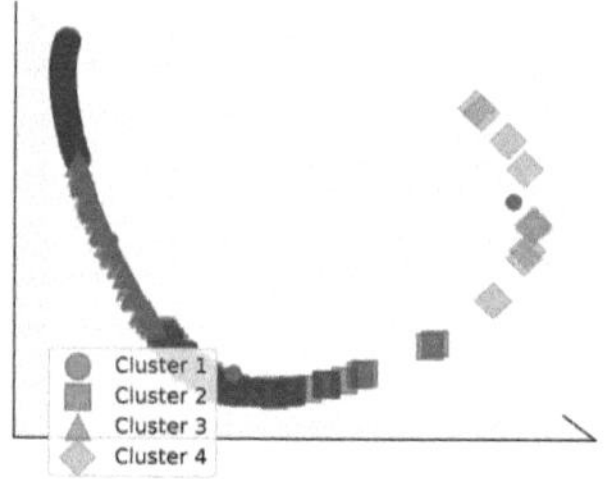

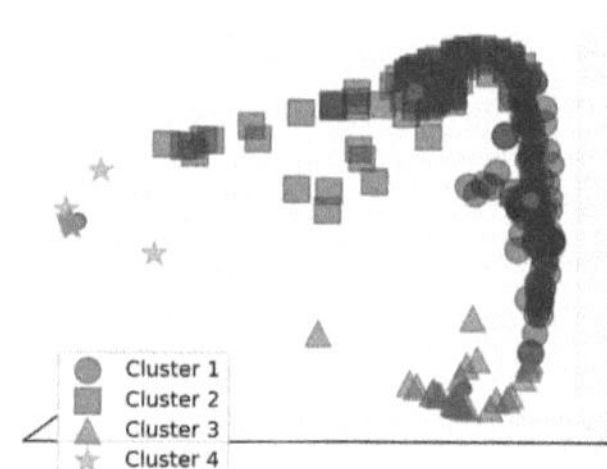

(a) Cluster labels of the 1D shape descriptor, applied on its corresponding spectral embedding

(b) Cluster labels of the 2D shape descriptor, applied on its corresponding spectral embedding

Fig. 3. Clusters determined by the two distinct shape descriptors

descriptors. By identifying the subject's identity who falls the closest to the cluster's centroid, we can identify the morphotype of the group, which could help us describe the shape of each population.

Extraction of Morphotypes. To explain the shape differences, in the case of both shape descriptors we have opted to use curve plots similar to the 2D shape descriptor, but instead of normalizing on both X and Y directions we only normalized the height of the curve plots, this way we were able to read circumferential differences of each determined cluster. This set of curve plots can be seen in Figs. 4b and 4d, wherein each cluster of the red curve represents the morphotype. The geometry of each morphotype is also visible in Figs. 4a and 4b. While the 1D shape descriptor uses the first 3 principal components of the 5 most variable zones of the lower limb (ankle at lateral malleolus (V1); ankle at its thinnest point (V2); mid lower leg (V4); calf (V7); tight (V16)), where the first PC represents the overall size growth of these zones, the second PC accentuates the ankle differences and the third PC highlights the tight growth, we can see these characteristics resonate in the curve plot clusters as well. The first cluster described by this shape descriptor has an arch at the lover leg region, coming from a smaller ankle/mid lower leg circumference, and has a well-developed calf region. From the curve plot group, we can also note that the upper tight region also presents a curve that can be a sign of a muscly tight. If we take a look at the morphotype of this group we can see that the quadriceps muscles are well developed and this resonates in its curve plot that shows an increased sudden growth at the lower tight region, which is approximately the head of vastus medialis and lateralis muscles. From the curve plots of Cluster 2, we can observe that the ankle region either increases or decreases in comparison to its morphotype and this group has the thickest ankle region from all the database, but the rest of the leg's circumferences decrease. The third cluster has a relatively small ankle and it is noticeable from both the morphotype and the curve plots that the shape of this group is conical or trapezoid, there is a smaller growth difference between adjacent circumferences, which results in a less curvy leg silhouette. The lack of the curve in the knee region compared to other morphotypes can be due to a padded knee or fat pad behind the knee. The fourth group is represented with a shape that grows both lower leg and tight region, it is the representation of a uniform fat deposit on the length of the leg. The morphotypes of the 2D shape descriptor show similarities to the ones determined by the 1D shape descriptor, but the results of these grouping are more shape-related rather than circumferential growth, which is since the 2D shape descriptor overlaps the curve plots and measures the pairwise distances, this way if two subjects are close in shape to each other, the clustering algorithm will group them. This can be visualized when the curve plots are represented with real circumferential values, the clusters are more uniform in shape. For example, the subjects in cluster 1 have the same shape of the lower leg and start to have differences in muscle development in the tight region. This group has a very similar morphotype as the 1st cluster of the 1D shape descriptor, an overall muscly leg. The second grouping is the trapezoid shape, this trapezoid shape is observable on the inner silhouette of the morphotype's leg, which lacks of arch in the lower knee and above ankle region. We can see this on the curve plot as the shape of this curve is straight, with minimal arch. The third group of the 2D shape descriptor is a leg shape with a lower muscle mass and the morphotype represents the smallest size from the group, and it can be seen on the curve plots that the size of this group uniformly grows, maintaining the same shape as the morphotype. The fourth group of the 2D SD is also similar to the fourth group of the 1D SD. These legs present a curve in the upper part of the thigh which is due to the inner tight and outer tight fat deposit. As the size of these subjects grows the tight and lower leg

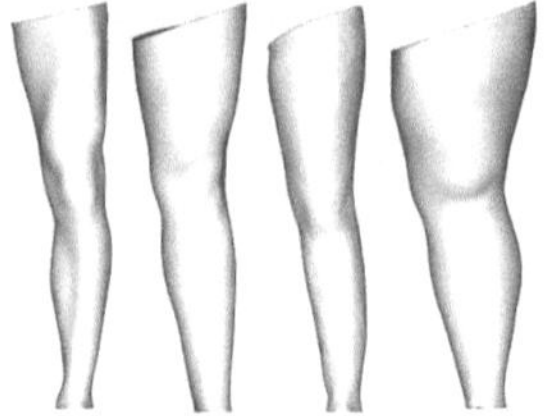

(a) Morphotypes of the 4 clusters determined by the 1D shape descriptors

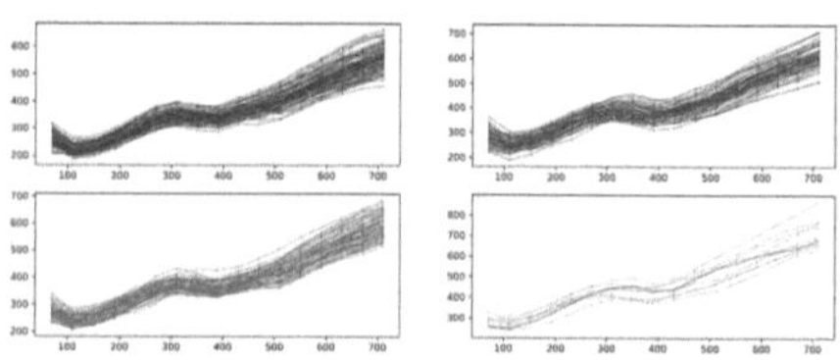

(b) Morphotypes of the 1D SD, represented with curve plots of the circumferences and the normalized heights of 17 cross sections along the leg length

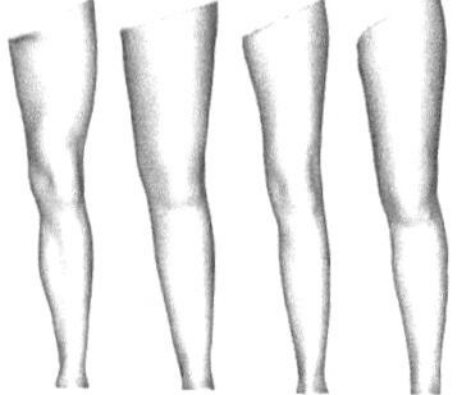

(c) Morphotypes of the four clusters determined by the 2D shape descriptors

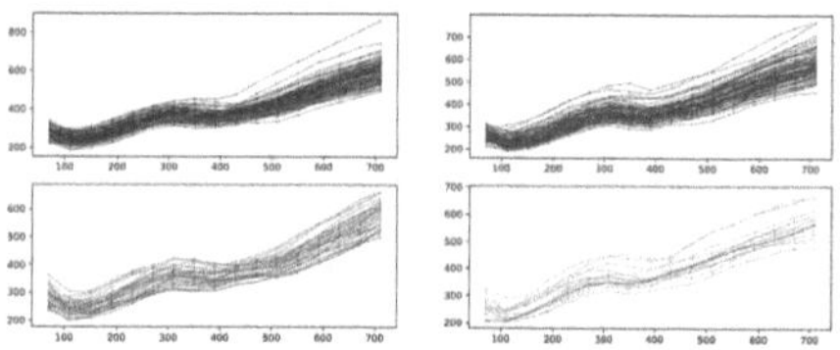

(d) Morphotypes of the 2D SD, represented with curve plots of the circumferences and the normalized heights of 17 cross sections along the leg length

Fig. 4. Extracted morphotypes

region grows significantly, but unlike the morphotype of the second group, these subjects maintain the inner knee "S" curve.

4   3D Lower Limb Shape Extraction

Data Acquisition 2nd Measurement Campaign. A second measurement campaign has been conducted in our research facility during which we have collected 3D human scans of 131 subjects, 74 of which are women (precisely 64 under 30 and 10 above the age of 30) and 57 men (50 under 30 and 7 above the age of 30). The position of the body is in a standard anthropometric standing position. The subjects have been asked to place 3D stickers on 10 anthropometric bony landmarks on their left legs. They have been instructed to palpate bone prominences of the following anthropometric landmarks, which are also illustrated in Fig. 5.: Ilium, Anterior Superior Iliac Spine (ASIS), Great Trochanter, Popliteal Fossa (fat pad), Lateral Epicondyle, Patella (kneecap), Medial Epicondyle, Tibial Tuberosity, Lateral Malleolus (outer ankle bone), Medial Malleolus (inner ankle bone). This operation was necessary as the body scanner (VITUSbodyscan®) with a precision of ± 1mm can detect the 3D stickers placed on the landmarks, therefore facilitating the data collection.

In our previous study, presented in Sect. 3, we have proven that 1D measurements, meaning circumferential measurements taken on the length of the

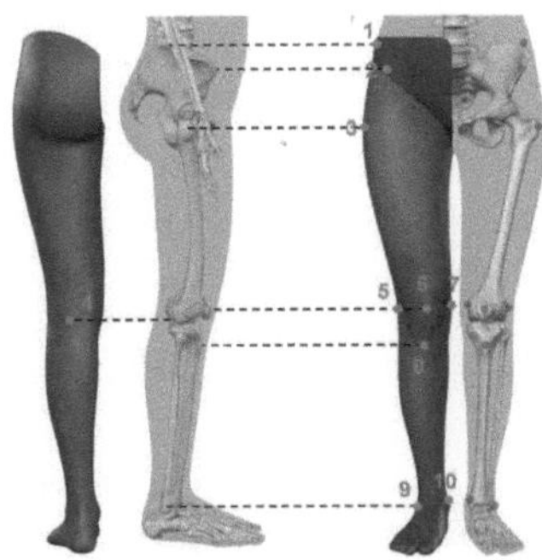

Fig. 5. Anthropometric soft tissue landmarks on the leg: 1.) Ilium, 2.) ASIS, 3.) Great Trochanter, 4.) Popliteal Fossa, 5.) Lateral Epicondyle, 6.) Patella, 7.) Medial Epicondyle, 8.) Tibial Tuberosity, 9.) Lateral Malleolus, 10.) Medial Malleolus

leg, can differentiate at some level the shapes of legs, creating 4 distinct cluster populations, but we have judged that let alone these measurements are not sufficient enough to precisely discriminate leg shapes. We have also introduced a 2D shape descriptor in the form of normalized curve plots illustrating the relationship between 17 circumferential measurements and the segment height (distance between two adjacent circumferences). This descriptor also gave valuable information about our data dividing it into 4 clusters. Comparing the labels of the two sets of clusters (determined by the 1D and 2D shape descriptors) we have found few correlations, deducting the conclusion that the shape of the leg needs to be further studied. Using the 10 anthropometric landmarks, we were able to automatically collect the 17 circumferential measurements along the leg length, we could determine a variation of the Q-angle (angle measured between the planes determined by the ASIS Patella and Patella Tibial Tuberosity) and the angle recurvatum (angle measured between the planes determined by the Great Trochanter Lateral Epicondyle and Lateral Epicondyle Lateral Malleolus. These two angles are not measured in the same way as clinically determined (with a standing position where the legs are straight), as in our case, the legs are slightly spread, which results in a slightly different angle in comparison to a straight-legged posture. The Q angle, visible in Fig. 6a., was taken into consideration because it can provide shape-related information, for example detecting conditions such as bowed leg or knocked knee [1, 20]. The recurvatum angle is visible in Fig. 6b, and it is considered for the same reason as the Q-angle.

Even if the clinical methods of measuring these two angles differ from our method, it can be viewed as reliable as all subjects have been instructed to stand in the same position, placing the landmarks on their legs with high attention and the measurements have been taken precisely by the software. Other measurements the software took were the length of the femur determined by the distance between the Great Trochanter and the Lateral Epicondyle and the length of the tibia determined by the distance between the Tibial Tuberosity and the medial malleolus. These two measurements provide us with proportion-

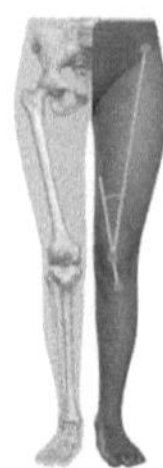

(a) Q-angle measured between the plane described by the ASIS and Patella and the plane described by the Patella and Tibial Tuberosity

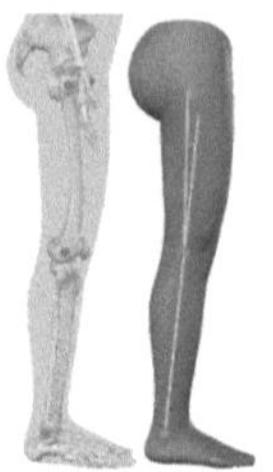

(b) Recurvatum angle between the plane by the Great Trochanter and Lateral Epicondyle and the plane by the Lateral Epicondyle and Lateral Malleolus

Fig. 6. Lower Limb angles

related information. As the software is capable of automatically determining the position of the crotch, gluteal fold, and calf, we were able to take into consideration these anthropometric landmarks as well. In addition, we have programmed the software to detect the extreme points of each circumference in X$\pm$ and Z$\pm$ directions. We took the XYZ cartesian coordinates of all the above-mentioned points and used them in future analysis.

3D Leg Shape Modeling and Shape Descriptor Extraction. Our challenge was to find a way to capture the shape differences of the leg using more than circumferential measurements. Our idea was to take longitudinal projections in different areas of the leg instead of cross sections. One of the challenges in doing so was to create a projectable polygon frame in a CAD software called DesignConcept 3D from Lectra, in which we can import one-by-one each scan, as well as the XYZ coordinates of the anthropometric points collected from Anthroscan®, and the frame would automatically adapt to the properties of the currently studied leg. In this section, the above-mentioned projection frame will be described, and the obtention of the output data from this software. Figure 7a Illustrates the 16 anthropometric points recovered from Anthroscan. The position of these points is manipulated with an Excel file that is linked to the predefined parameters of the cartesian points in DC3, meaning that if the subject changes, the position of the points changes as well. Figure 7b illustrates the extreme points on the 15 cross sections (17- crotch and ankle) in X$\pm$ and Z$\pm$ direction. This figure also illustrates that if we connect longitudinally these points on the four sides of the leg, we can visualize a wavy display of the respective lines.

We wanted to achieve a level of linearity when we connected these points. Therefore, we chose four anthropometric points on each side of the leg and connected each adjacent point with a straight line. The goal was to align the extreme points that fall between two adjacent landmarks in Fig. 8 following the rule of linear interpolation between two known points. We import the x,y or y,z

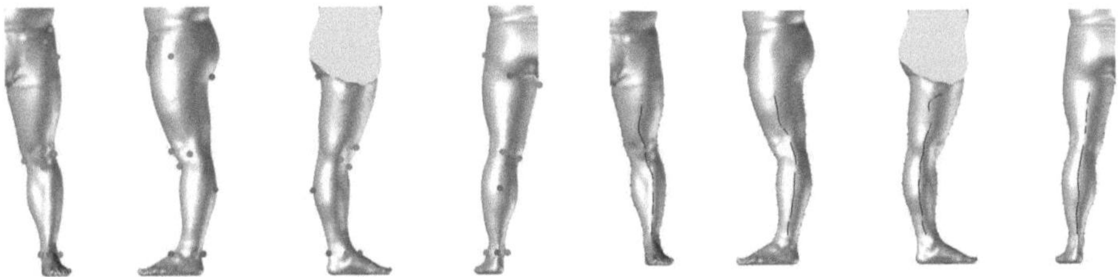

(a) Anthropometric landmarks exported from Antroscan® imported in DesignConcept® 3D.

(b) Extreme points in X+/- and Z+/- direction on the 15 circumferences connected by a spline.

Fig. 7. Transfer of the Anthroscan landmarks in a CAD software

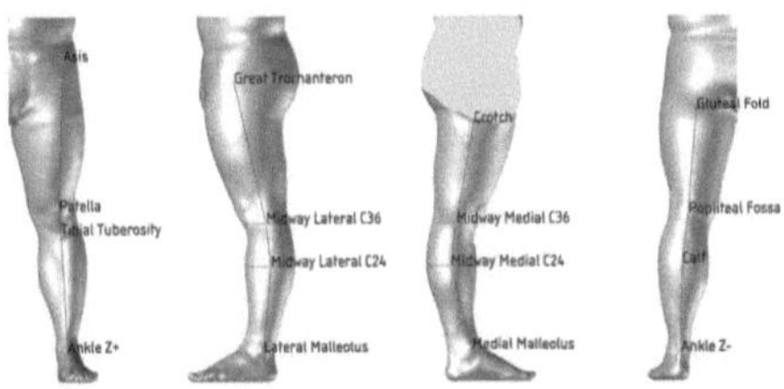

Fig. 8. Chosen landmarks used as the "known points" in the linear interpolation Eq. 1

coordinates of the original extreme point (depending on which side of the leg the point is situated), then we calculate the third coordinate with the help of Eq. 1 (Fig. 9). The resulting points, connected by a spline, can be seen in Fig. 10a. Figure 10b illustrates the original spline (black) and the new spline (green).

$$x = x_1 + \frac{(y - y_1)}{(y_2 - y_1)} \times (x_2 - x_1) \tag{1}$$

The role of the new point configuration was to take the silhouette of the leg from 4 sides in a controlled way, considering as many anthropometric points as possible. The inner and outer silhouettes can also be compared to side seams on form-fitted pants. These silhouettes also represent the alignment of the leg and the Q-angle, which is a comparability factor between individual legs.

This realignment operation was important from the point of view of comparability because if we leave the extreme point in its original state, this gives an uncontrolled spline for every subject, as the leg soft tissue is unique in every subject's case, making the original spline unusable.

After having the optimal position of the above-mentioned points, we continued by creating a framework on these points; in this way, if all our frame is connected to parametrized points, once the parameters change, the frame changes with them. The idea was to get the leg silhouettes from as many sides as possible. To create this frame, we chose 5 levels on the leg. 1st level is at the ankle, 2nd level is at level C12 (approximate bronchial level), 3rd level is a com-

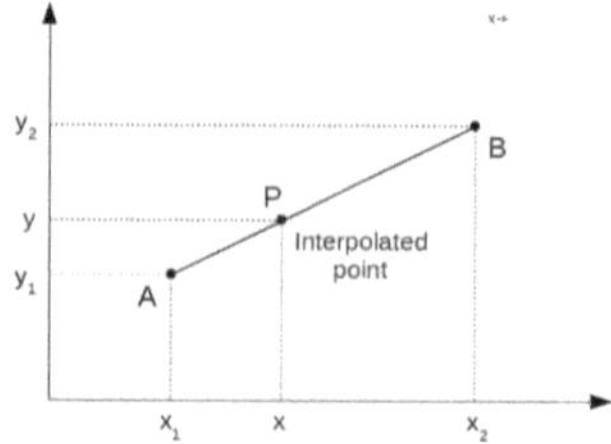

Fig. 9. Linear interpolation between two known points, A, representing a landmark on Fig. 8 and B, representing the adjacent pair of A; P is the extreme point that falls between A and B. Each extreme point has two known coordinates and the third is calculated with formula

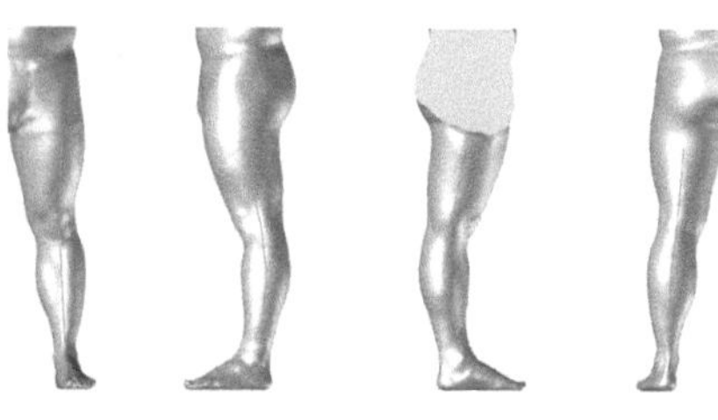

(a) New position of extreme points and the corresponding spline that connects them. In this state, the position of the points is not necessarily on the surface of the leg as they have been shifted in X and Z direction.

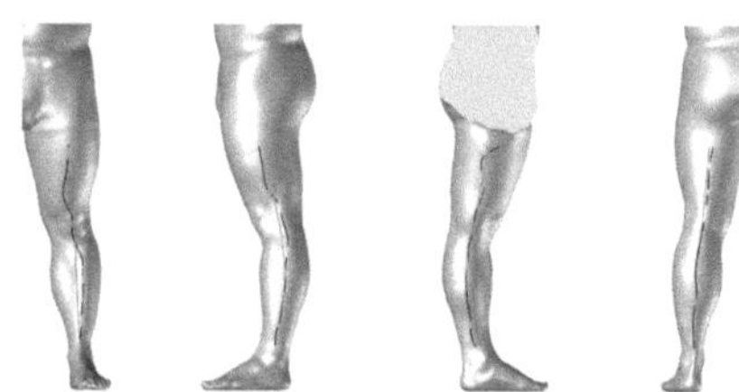

(b) Comparison between the original state of the extreme points and the new position of points that are not extreme points anymore but are positioned in a controlled way, same for every subject.

Fig. 10. Result of the point interpolation method (Color figure online)

bination between level C24 (approximate calf level) and the Tibial Tuberosity, the 4th is at level C36 (approximate knee level) and the 5th is also a combination between the crotch, ASIS projection (Fig. 11a: yellow line between ASIS and Patella, the red point on yellow line representing the intersection point of the plane between the crotch and Trochanter and the yellow line. This red point is projected on the surface of the leg.), Great Trochanter and the Gluteal Fold. We connected the corresponding four points at each level, which can be visualised also in Fig. 11a. With 3D transformations, we translated the points on the outside of the leg surface to stay on a line connecting the barycentre of the corresponding level and the translated point. This way, we have created contours connecting the translated points that will be projected on the surface of the leg. The first set of projection contours that create the silhouette of the leg on the four sides (left, front, right, back) can be visualized in Fig. 11b and the resulting projected lines in Fig. 11c These four projection contours are automatically adapting to the subject in study, once the XYZ coordinates are imported, meaning no further manual manipulation is needed.

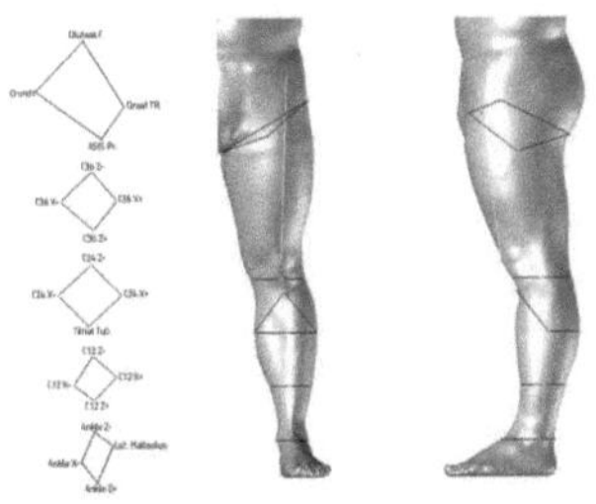 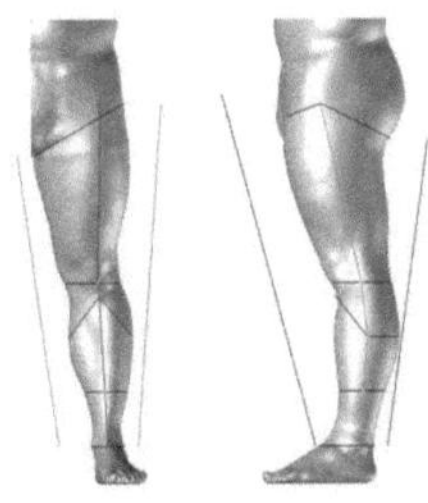

(a) The five levels on the leg serving as a basis for the projection frame;ASIS projection method

(b) Projection lines that give the silhouette from each side of the leg (left, front, right, back)

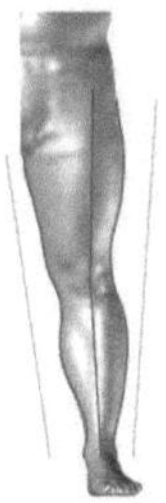

(c) Projected lines on the surface of the leg obtained with the help of the projection lines

Fig. 11. Creation process of the first set of projection lines (Color figure online)

A second set of projection contours has been created to get the projection of the leg midway between the 4 side silhouettes. We repeated the same transformation of midway points visible on the right of Fig. 12a to the outside of the leg and created projection contours in the form of a 3D spline. We have created a plane between levels C36 and C24 on each projection contour, representing the projection plane and serving as director of projection. Because we are dealing with a 3D spline that is not placed in the same plane at every level (the ankle, bronchial, and crotch are not in the same plane as the knee and calf) therefore, the projection of these points would not fall in the correct place. To remediate this problem, we have introduced sliding points at the level of the crotch, bronchial, and ankle, that will approach the projection contour to the surface of the leg, this way assuring the correct position of projection. Because the knee and calf levels are always projected correctly (as they are in the same plane as the projection plane), we have receded the contour from the leg surface, so we can ensure that the contour stays outside of the leg surface at all levels. For the reason of the sliding points, before projecting the contours on the leg's surface, manual operation of approaching these points had to be introduced.

Similarly, to the midway contours, we have created a third set of projection contours, this time targeting the mid portion between a red projection line (sil-

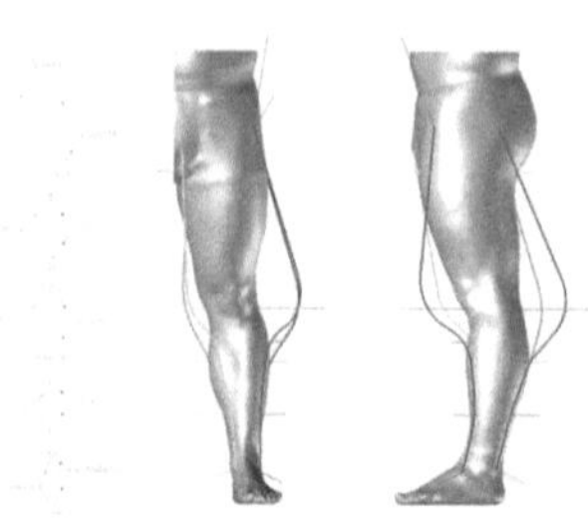

(a) Midway points between the silhouettes that are translated out of the leg

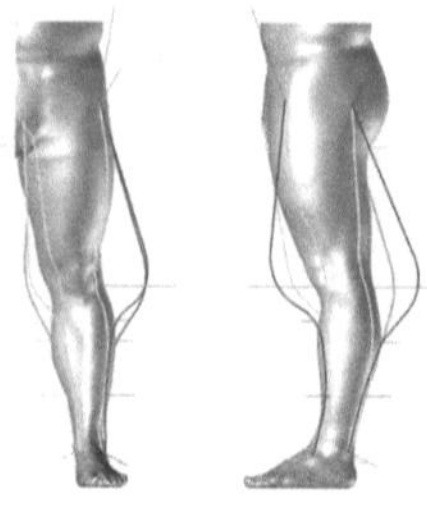

(b) Projection frame and projected contours midway between the silhouettes.

Fig. 12. Creation process of the second set of projection lines

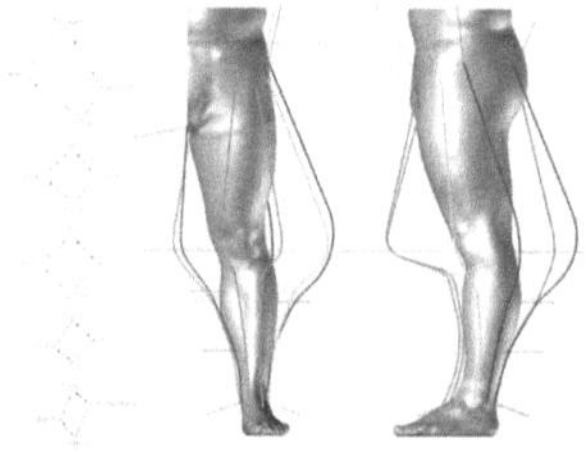

(a) Points between the silhouette points and mid-points creating the third set of projection frame

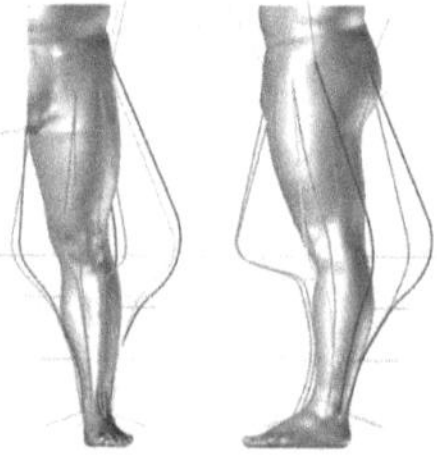

(b) Third set of projected lines, that along with the others divide the leg in 16 non-equal segments

Fig. 13. Creation process of the third set of projected lines (Color figure online)

houette, Fig. 13.) adjacent to a yellow projection line (midway, Fig. 12b). The process is identic to the midway projection contours and the shape of the third set of points and projection contours can be visualized in Fig. 13a The resulting projected lines in Fig. 13b divide the leg's circumference into 16 non-equal segments and this characteristic of the projection is perfectly repeatable on any leg shape.

We limited the projected lines to three steps to define the leg portion that will serve as input in the classification algorithm (Fig. 14). The first step is a limitation of the upper frontal barrier, which happens with the help of a plane defined between the crotch point and the Great Trochanter; all lines falling between these two landmarks have been limited with this plane. The second step comprises the limitation of the lower leg barrier, which also happened with the help of a horizontal plane defined at the level of the Lateral Malleolus. At this step, all of the projection lines are limited. The third step is a limitation of the upper back-barrier, which happens with the help of sliding points placed

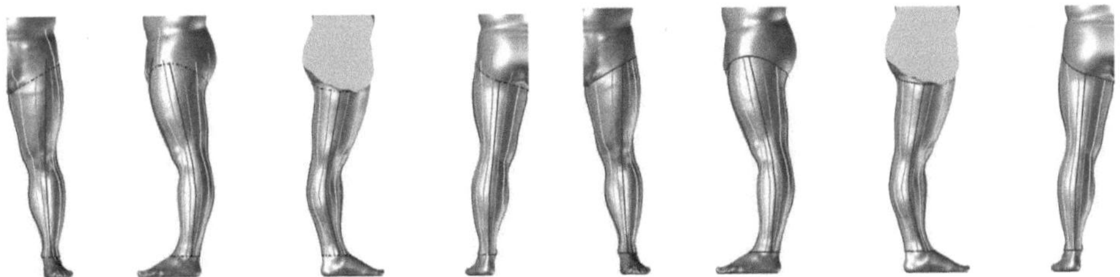

Fig. 14. Limitation of the projected lines with planes and points.

Fig. 15. Creation process of the third set of projection lines

on the projection lines to contour the shape of the buttocks. This zone is highly variable, with the highest difference between men and women; therefore, finding the position of these sliding points is a manual operation. When all lines have been limited, a contour spline has been interpolated through each endpoint, at both the upper and lower end of the leg. The interpolation frame can be visualized in Fig. 15.

It has to be noted that we have chosen to use this specific contouring of the upper barrier of the leg instead of simply cutting the leg with a horizontal plane at the crotch level because the lowest point of the glutes sometimes falls under the level of the crotch, especially in case of women, therefor we judged that this way the shape will also contain information about this particular area, conserving the shape more precisely.

Using the projection line frame, we reconstructed the leg shape with a patchwork function and made the leg a watertight manifold (solid), visible in Fig. 16. On this watertight leg, we created mesh regions with variable triangle size: 10mm edge triangulation (Fig. 16b), 15mm edge triangulation (Fig. 16c), 25mm edge triangulation (Fig. 16d), resulting in a different number of faces, which is important in the classification algorithm. Having three different meshes of the same shape is useful from the point of view of data augmentation, tripling the size of our data as we want our model to learn tessellation-invariant features. In addition to the meshes, we also extracted the .dxf file of the 16 projection lines (Fig. 16e) and the .dxf file of 48 circumferences (Fig. 16f). These two files have been further modified in software called WeCov3r Studio, where we could extract the xyz cartesian coordinates of the polygons. In the case of the 16 projection lines, we extracted 50 equally distance points on each polygon, and in the case of the 48 circumferences, we extracted 48 points on each circumference. Figure 17 is an example of five different leg shapes with 16 projected lines that could be considered a simplified representation of the leg's shape. These projected lines can describe the shape of the leg by creating a patchwork surface or b-spline interpolated surface with the help of these lines we have proven that it gives back the real shape of the leg or a shape that is very close to reality.

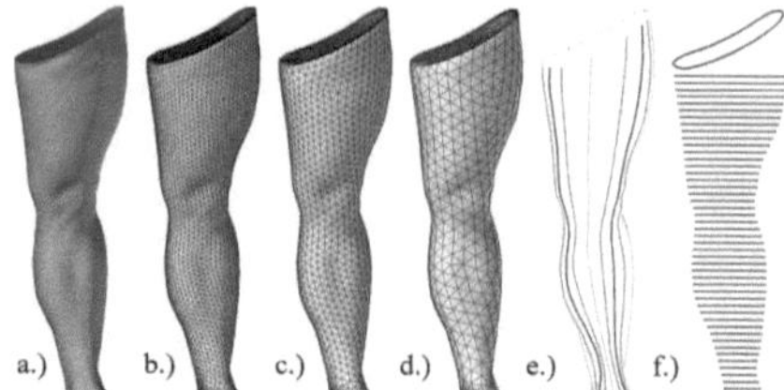

Fig. 16. Final data exported from DesignConcept: a.) Patchwork watertight 2-manifold, b.) 3D mesh with 10mm edge triangle, c.) 3D mesh with 15mm edge triangle, d.) 3D mesh with 25mm edge triangle, e.) .dxf format of the projected lines, f.) .dxf format of the 48 circumferences

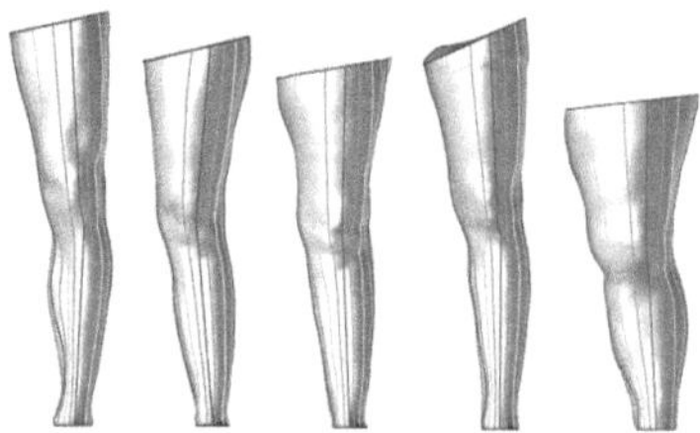

Fig. 17. Projected lines on different shapes of legs

5   Difficulties, Challenges, and Perspectives

This section highlights the key difficulties and challenges faced during our research and the upcoming perspectives.

Difficulties and Challenges about Deep Learning from 3D Meshes
One of the primary difficulties encountered in this study was the prolonged pre treatment of data as the shape descriptor extraction and leg segmentation in a CAD environment took a longer time. The database preparation represented a challenge as this process cannot be fully automated. To address this issue, we are developing CAD software that could significantly speed up the pre-treatment process. Another difficulty was the scarcity of data, as our database is composed of 131 subjects, so implementing a deep learning method on this data is rather challenging. Still, the CNN algorithm that we will use is not sensible to tessellation invariance, meaning that the correspondence between shapes remains consistent even if the triangulation is different. This way we can augment our data to up to 2751 subjects with random rotations and remeshing of the leg shapes. Additionally, we will collaborate with a Hospital in France, which will allow us to collect the 3D surface scans of lymphoedema patients, which will allow us in first hand to enlarge our database and in second hand to be more aware of shape changes that this pathology could cause as well as to see if the classification algorithm will be able to determine distinct morphological groups.

Perspectives about Deep Learning from 3D Meshes

Our mesh data, which we have recovered from the surface reconstruction in Sect. 4 is presented as non-Euclidean data, which arises from its irregular structure, intrinsic geometric properties, and the need for specialized computational methods to process it effectively. The surface of a 3D mesh is a 2-dimensional manifold embedded in 3-dimensional space, and its local properties (like curvature) are non-Euclidean. In our study, we obtained a database of scanned human subjects, presented at first in the form of point clouds transformed into polygonal meshes that have undergone a pre-treatment process of smoothing, hole-filling, and noise removal. Therefore, they can be considered in their initial form as non-Euclidean data. We would like to preserve the topological information that the triangular meshes could provide; therefore, we will opt for deep learning methods to process triangular meshes as input. SubdivNet [12], takes as input closed 2-manifold triangle meshes with subdivision sequence connectivity, extracts a 13-dimensional vector (7-d shape descriptor: the face area, the three interior angles of the triangle, and the inner products of the face normal with the three vertex normals; 6-d pose descriptor: position of the face center and the face normal). They have designed a mesh convolution operator that can support variable kernel size, dilation, and stride, just as in the image domain as well as a mesh pooling and upsampling operator. This is important because the preserved feature vectors can be introduced in successful 2D CNN algorithms such as ResNet50 [11] or Deeplab [5]. The advantage of both MeshCNN and SubdivNet is that they work with relative features, making them invariant to rotation, translation, and uniform scaling.. [10]. To prepare the data a subdivision loop connectivity remeshing was necessary on our 3D leg shapes. For this, we have applied the MAPS algorithm [18] on the segmented leg exported from the CAD software we used, with around 8122 faces. With the help of this input mesh, we want to extract latent space feature vectors, which are compressed information about the geometry of the 3D leg meshes. These feature vectors can then be further analyzed with unsupervised clustering methods to see the best partition in the data, resulting in clusters that we translate as morphological groups within our population.

6   Discussion and Conclusion

In this study, we aimed to improve the design and effectiveness of medical compression stockings (MCSs) by using 3D body scans to extract morphotypes. This research is particularly relevant for treating chronic venous insufficiency (CVI), a common disease that affects a significant portion of the global population. Our work is part of the broader Symphonies initiative, which focuses on digitizing the creation process of MCSs to enhance their fit and therapeutic efficacy.

Our methodology involved collecting a 3D database of human leg scans, along with anthropometric measurements and landmarks, which together have been used to extract shape and size-related information about the leg geometry. These data were first introduced in classic unsupervised clustering algorithms, such as

Spectral clustering and K-means clustering, defining 4 distinct clusters for both 1D and 2D data. The results of these classifications allowed us to identify the four main shapes of human legs, and to delve even deeper into the shape analysis we identified longitudinal sections on the leg, that have the potential to simplify its geometry. Our future objective is to prove that these longitudinal sections combined with the 3D meshes can describe the shape of the leg better than 1D and 2D data.

One of the significant findings of our research is the importance of personalized MCSs. Traditional sizing systems often fail to account for individual variations in leg shape, leading to poor fit and reduced patient compliance. The idea of a morphotype that can describe the leg shape of a specific population would address this issue by providing a more accurate representation of leg shapes, thereby improving the comfort and effectiveness of MCSs. By creating an adaptive parametric model from each of the found morphotypes we would provide an accurate sizing reference form for the industry of medical compression stocking, that they could use as a 3D support in pattern creation.

However, the study also highlighted several challenges. Previous studies proved that one of the major difficulties is the high non-compliance rate among patients using compression products. Factors such as skin irritation, discomfort, and difficulty in donning the stockings contribute to this problem. Our research suggests that personalized MCSs, tailored to individual leg shapes, can significantly enhance patient compliance by improving fit and comfort. The concept of personalizing these products is still considered a great challenge in an industrial setting as it would reduce considerably the production time. For this, the optimal solution would be to create parametric models from the proposed morphotypes and assign to each morphotype a new sizing system, so that the industry would have off-the-shelf products adapted to a greater population.

Another challenge is the complexity of accurately capturing and classifying 3D leg shapes. While our deep learning methods have shown promise, there is still room for improvement in computational efficiency and accuracy. Future work will focus on refining these algorithms and expanding our dataset to include more diverse populations, particularly older individuals and those with lymphoedema.

In conclusion, our research has demonstrated the potential of using 3D body scans and deep learning algorithms to develop personalized MCSs. By creating adaptive morphotypes that accurately represent individual leg shapes, we can improve the fit and effectiveness of compression therapy for CVI patients. Future directions will involve optimizing our algorithms, expanding our dataset, and collaborating with medical professionals to ensure that our products meet the needs of all patients. This innovative approach holds promise for significantly enhancing the quality of life for individuals suffering from CVI and related conditions.

References

1. Q angle. https://www.physio-pedia.com/Q_Angle Accessed 11 May 2024
2. How to measure compression products (2024). https://www.sigvaris.com/en-ca/expertise/how-to/measure
3. Banfalvi, T., Bruniaux, P., Tartare, G., Dassonville, F.: Unsupervised classification of 3D morphologies of human legs applicable in the use of medical compression stocking design. In: Proceedings of the 18th "Ergonomie et Informatique Avancée" Conference, pp. 1–8 (2023)
4. Berti-Hearn, L., Elliott, B.: Chronic venous insufficiency (2019)
5. Chen, L.C., Papandreou, G., Kokkinos, I., Murphy, K., Yuille, A.L.: DeepLab: semantic image segmentation with deep convolutional nets, Atrous convolution, and fully connected CRFs. IEEE Trans. Pattern Anal. Mach. Intell. **40**(4), 834–848 (2017)
6. Davies, A.H.: The seriousness of chronic venous disease: a review of real-world evidence. Adv. Ther. **36**(Suppl 1), 5–12 (2019)
7. Eberhardt, R.T., Raffetto, J.D.: Chronic venous insufficiency. Circulation **130**(4), 333–346 (2014). https://doi.org/10.1161/CIRCULATIONAHA.113.006898
8. Erickson, C.A., et al.: Healing of venous ulcers in an ambulatory care program: the roles of chronic venous insufficiency and patient compliance. J. Vasc. Surg. **22**(5), 629–636 (1995). https://doi.org/10.1016/S0741-5214(95)70051-X
9. Franks, P.J., Oldroyd, M.I., Dickson, D., Sharp, E.J., Moffatt, C.J.: Risk factors for leg ulcer recurrence: a randomized trial of two types of compression stocking. Age Ageing **24**(6), 490–494 (1995). https://doi.org/10.1093/ageing/24.6.490
10. Hanocka, R., et al.: MeshCNN: a network with an edge. ACM Trans. Graph. (ToG) **38**(4), 1–12 (2019)
11. He, K., Zhang, X., Ren, S., Sun, J.: Deep residual learning for image recognition. In: Proceedings of the IEEE conference on computer vision and pattern recognition, pp. 770–778 (2016)
12. Hu, S.M., et al.: Subdivision-based mesh convolution networks. ACM Trans. Graph. (TOG) **41**(3), 1–16 (2022)
13. Ilska, A., et al.: Using a 3D body scanner in designing compression products supporting external treatment. Fibres and Textiles in Eastern Europe **Nr 5 (125)** (2017). https://doi.org/10.5604/01.3001.0010.4636
14. Jindal, R., Uhl, J.F., Benigni, J.: Sizing of medical below-knee compression stockings in an Indian population: a major risk factor for non-compliance. Phlebology **35**(2), 110–114 (2020). https://doi.org/10.1177/0268355519854611
15. Jolliffe, I.T.: Principal Component Analysis, 2nd edn. Springer Series in Statistics, Springer, New York (2002)
16. Kankam, H.K.N., Lim, C.S., Fiorentino, F., Davies, A.H., Gohel, M.S.: A summation analysis of compliance and complications of compression hosiery for patients with chronic venous disease or post-thrombotic syndrome. Eur. J. Vasc. Endovasc. Surg. **55**(3), 406–416 (2018). https://doi.org/10.1016/j.ejvs.2017.11.025
17. Karakashian, K., Shaban, L., Pike, C., van Loon, R.: Investigation of shape with patients suffering from unilateral lymphoedema. Ann. Biomed. Eng. **46**, 108–121 (2018)
18. Lee, A.W., Sweldens, W., Schröder, P., Cowsar, L., Dobkin, D.: Maps: multiresolution adaptive parameterization of surfaces. In: Proceedings of the 25th annual conference on Computer graphics and interactive techniques, pp. 95–104 (1998)

19. Liu, R., et al.: Stratified body shape-driven sizing system via three-dimensional digital anthropometry for compression textiles of lower extremities. Text. Res. J. **88**(18), 2055–2075 (2018). https://doi.org/10.1177/0040517517715094
20. Liu, R.: Stratified body shape-driven sizing system via three-dimensional digital anthropometry for compression textiles of lower extremities. Text. Res. J. **88**(18), 2055–2075 (2018)
21. Lu, Y., Yang, Z., Wang, Y.: A critical review on the three-dimensional finite element modelling of the compression therapy for chronic venous insufficiency. Proc. Inst. Mech. Eng. H **233**(11), 1089–1099 (2019). https://doi.org/10.1177/0954411919865385
22. Lurie, F., Bittar, S., Kasper, G.: Optimal compression therapy and wound care for venous ulcers. Surg. Clin. North Am. **98**(2), 349–360 (2018). https://doi.org/10.1016/j.suc.2017.11.006
23. Lurie, F., Bittar, S., Kasper, G.: Optimal compression therapy and wound care for venous ulcers. Surg. Clin. **98**(2), 349–360 (2018)
24. von Luxburg, U.: A tutorial on spectral clustering (2007). http://arxiv.org/abs/0711.0189 Accessed 08 Mar 2023
25. Nørregaard, S., Bermark, S., Gottrup, F.: Do ready-made compression stockings fit the anatomy of the venous leg ulcer patient? J. Wound Care **23**(3), 128–135 (2014). https://doi.org/10.12968/jowc.2014.23.3.128
26. Partsch, H.: Mechanism and effects of compression therapy. In: The Vein Book, pp. 103–109. Elsevier (2007). https://doi.org/10.1016/B978-012369515-4/50013-2
27. Partsch, H.: Mechanism and effects of compression therapy. In: The vein book, pp. 103–109. Elsevier (2007)
28. Rehman, Z.U.: Pattern of chronic venous insufficiency among patients presenting to a vascular surgery clinic in low- to middle-income countries (LMIC): a cross-sectional study. Ann. Vasc. Dis. **14**(2), 118–121 (2021). https://doi.org/10.3400/avd.oa.20-00088
29. Reich-Schupke, S., Gahr, M., Altmeyer, P., Stücker, M.: Resting pressure exerted by round knitted moderate-compression stockings on the lower leg in clinical practice–results of an experimental study. Dermatol. Surg. **35**(12), 1989–1998 (2009). https://doi.org/10.1111/j.1524-4725.2009.01318.x
30. Rotsch, C., Oschatz, H., Schwabe, D., Weiser, M., Möhring, U.: Medical bandages and stockings with enhanced patient acceptance. In: Handbook of Medical Textiles, pp. 481–504. Elsevier (2011). https://doi.org/10.1533/9780857093691.4.481
31. Rousseeuw, P.J.: Silhouettes: a graphical aid to the interpretation and validation of cluster analysis. J. Comput. Appl. Math. **20**, 53–65 (1987). https://doi.org/10.1016/0377-0427(87)90125-7
32. Tan, M., Urbanek, T., Rabe, E., Gianesini, S., Parsi, K., Davies, A.H.: Compression therapy in the management of varicose veins. Phlebology **39**(4), 276–279 (2024). https://doi.org/10.1177/02683555231222679
33. Tandler, S.F.: Challenges faced by healthcare professionals in the provision of compression hosiery to enhance compliance in the prevention of venous leg ulceration. EWMA J., 29–33 (2016)
34. Xi, W., Bao, Y., Qiao, L., Xia, G., Xiaoming, T.: Parametric modeling the human calves for evaluation and design of medical compression stockings. Comput. Methods Programs Biomed. **194**, 105515 (2020). https://doi.org/10.1016/j.cmpb.2020.105515

Application of Digital Survey Tools to Screen for Child Maltreatment Among Secondary School Students in Vietnam

Thi Minh Le[1]([⊠]) [iD], Tho Nguyen[1,2], Ha Dinh Thu[1] [iD], Bui Thi Phuong[1] [iD], and Huong Thanh Nguyen[1] [iD]

[1] Hanoi University of Public Health, 1A Duc Thang, Bac Tu Liem, Hanoi, Vietnam
lmt@huph.edu.vn
[2] Dong A University, Danang, Vietnam

Abstract. Effective data collection and management are vital for research, and digital tools are increasingly used to streamline processes and reduce errors. For a study on child maltreatment in Vietnam, we selected a digital survey tool to manage and visualize sensitive data. We conducted four phases: document review; testing KoboToolbox for lean operation, skip logic and validation criteria; a pilot survey; and a main survey with 2326 students aged 12–15 at eight schools in Bac Giang province. We compared digital survey tools and paper-based platforms to choose the best method for screening child maltreatment. KoboToolbox proved convenient, time-saving, and free, with validation features to reduce data bias and a user-friendly interface for both experts and students. Its offline capabilities addressed connectivity issues in local settings. KoboToolbox was selected for its ability to construct sensitive questionnaires, enhance data collection, and improve quality, contributing to a better understanding of child maltreatment in Vietnam.

Keywords: KoboToolbox · digital survey tool · online platform · child maltreatment · data collection · data management · Vietnam

1 Introduction

Data collection apps or tools are applications that allow researchers to gather information, build forms, and create surveys to collect primary data. With these apps, a variety of methods, including computer-assisted phone interviewing, in-person interviews, or self-administered surveys [1]. There are some digital survey tools such as KoboToolbox, Unipark, and Qualtrics (DSTs) used in mental health, psychological, or child maltreatment studies [1, 2]. These DSTs support optimizing data collection, management, and analysis, and investigators and respondents choose their DSTs based on purposes and DSTs' characteristics [3]. Exploiting outstanding features of DSTs for choosing a survey tool is crucial for optimizing resources and enhancing the quality of data collection results.

S. Thomassey et al. (Eds.): RAIDS 2024, LNICST 673, pp. 35–47, 2026.
https://doi.org/10.1007/978-3-032-14055-5_3

For maltreatment research, all maltreatment dimensions – physical, emotional, sexual abuse, and neglect must be collected, and exploring associated factors including the way of disclosure of child maltreatment information is essential [4]. Thus, DSTs are preferred by researchers in constructing quickly and conveniently long, complex, and sensitive questionnaires, for example, good skip logic questions. Another imperative feature of advanced survey tools is that researchers can read reports through data visualization in real-time during the survey [5–7]. In fact, for studies that survey a large sample size of the population, if pilots are not conducted to check the questionnaires' accuracy, many unexpected problems may occur in the central survey, leading to low-quality data and results [8]. Utilizing DSTs saves resources and more reliable results [9, 10].

Child maltreatment remains persistent in low-to-middle and high-income countries [4]. Maltreatment has many different styles and related factors like an information matrix of social networks (i.e., family, friends,…) that may affect each aspect of maltreatment. Thus, designed questionnaires are extraordinarily complex and lengthy, with many skip logic or cross-checked questions. To address issues, using DSTs can reduce confusion by linking questions for objects (students at secondary schools) and lower the time-consuming of rechecking previous questions while completing the questionnaire. In this paper, we discuss why the KoboToolbox app is selected for data survey of child maltreatment studies in Vietnam on some aspects: suitable and convenient for studies having large sample population, user-friendly interface suitable for the student, may easily design and save time when constructing a questionnaire, ensure security and privacy, or which one is valuable for international and domestic collaboration (multicenter studies) in maltreatment studies.

Description of the Study

This paper is a part of the project entitled "The role of schools in detecting and responding to child maltreatment: A collaborative approach to evidence-based policy development in Switzerland and Vietnam". Research on child maltreatment has shown that victims are at strongly elevated risks of negative short-term and long-term developmental outcomes [4]. A major obstacle in the effort to implement effective interventions against child maltreatment is that many victims of child maltreatment never disclose their experience to anyone, which makes it less likely that they will ever be addressed by a targeted response.

This paper is to discuss the DSTs to collect sensitive child maltreatment data via a three-phase preparation, then the researcher's experience in the pilot phase with 241 participants and in the main survey with 2326 students. In this article, we present the advantages and disadvantages of four types of data collection including KoboToolbox, Unipark, Qualtrics and paper-based methods. In addition, the paper describes our experiences in three phases of the study and discusses which DST to screen for child maltreatment in the context of Vietnam.

2 Methods

2.1 Study Design

There are three phases: In the first phase, we reviewed documents related operation, and functions of DSTs/paper-based method and their application capability in maltreatment studies. In preparation for data collection, four experts tested questionnaires, conditional questions, validation criteria, friendly user interface, time-consuming for completing a survey. We designed the questionnaire in 3 platforms for the pilot: paper-based, Unipark and KoboToolbox.

In the second phase, we tested the questionaire on paper-based and digital platforms. The digital platform was selected after pilot 15 students used each platform: paper-based survey, Unipark, and Kobotoolbox in Hanoi. The KoboToolbox was selected to pilot in the field site. A pilot study with 241 for students a secondary school was conducted using tablets, then researchers revised the tools and process of main data collection.

In the third phase, we conducted data survey using KoboToolbox platform in 8 secondary schools (4 schools in rural area and 4 schools in town) in Hiep Hoa district, Bac Giang province. 2326 students completed the questionnaire using tablets at schools.

2.2 Questionnaire

The questionnaire had 11 parts with 480 questions with 296 questions needing skip logic and three questions having validation criteria. There are 996 field data that need to be installed in the apps. Data collection was conducted from Feb, 2023 to December 2023.

2.3 Participants

In the pilot survey, a total of 241 students (including students from three classes of 7th grade and three classes of 9th grade) at a secondary school in Hiep Hoa district, Bac Giang province involved in the study. The students filled out the questionnaire in class using tablets anonymously. We rented 50 tablets and each time, one class (about 40 students) per time.

In the main survey, we conducted data collection involving adolescents between 13 and 15 years old from eight secondary schools in Bac Giang province with the expected sample size is 2200 students. The study design is to measure child maltreatment and disclosure of maltreatment. In reality, we collected 2356 records and after cleaning, the sample size for data analysis is 2326 students. It took about 30–65 min per student to complete the questionnaire.

2.4 Assessment Criteria

Features of KoboToolbox are evaluated such as platform, survey design, data collection, data security, management and analysis, and time-consuming finishing survey.

KoboToolbox software architecture (Fig. 1):

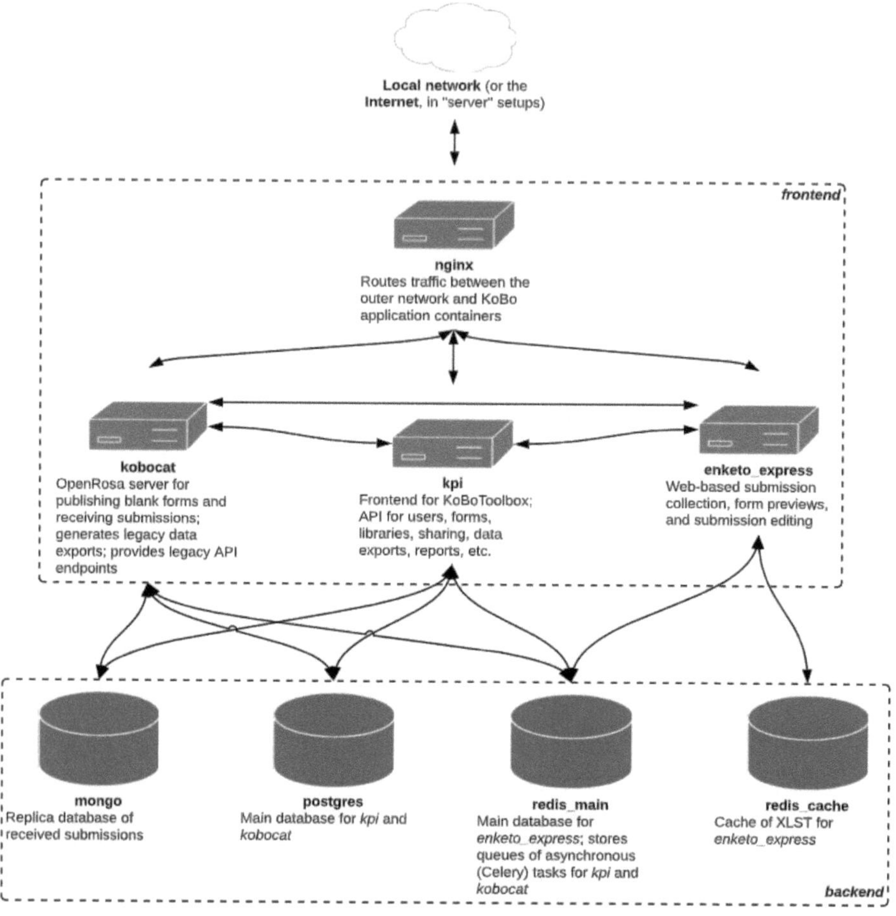

Fig. 1. High level server-side overview of KoboToolbox architecture *(source: Software Architecture)*

- kpi - for creating survey forms and reusing assets through a question library.
- kobocat and kobocat templates - for deploying surveys, collecting, and analyzing data.
- enketo-express - HTML5 Web app for collecting data, previewing forms, and editing data submissions.
- Kobocollect - Android app for collecting data.

KoboToolbox is based on elements from several other open source tools, most importantly pyxform, formhub/onadata, and the OpenDataKit. As a result, all forms and collected data are compatible with these tools.

2.5   Ethical Issue

This study was approved by the IRB of Hanoi University of Public Health 022-3211DD -YTCC dated 11 July 2022.

3   Results

3.1   Description of the Digital Survey Platforms

KoboToolbox is DST that allows researchers to collect, and manage data electronically. Through these tools, respondents can approach form interfaces more intuitively and conveniently than papers. They are widely used by many organizations such as humanitarian action, environment, human rights, public health institutes, research organizations, and education facilities and are applied in various fields. These tools are useful for multinational studies because of their multilingual translation (hundreds of languages), making it easy to work with global teams. KoboToolbox was founded in 2005. It is a free and open-source platform for the collection, management, and visualization of data [6]. KoboToolbox supports collecting online and offline data, when the Internet connection is available offline data will be synchronized in the online system.

Unipark, founded in 2004 by Unipark Software GmbH, is a popular German survey software widely used for academic research, market research, and data collection. It offers customizable survey creation, comprehensive plausibility checks, and automated testing tools, ensuring high-quality data [7]. While it has a user-friendly interface, respondents need an internet connection to access the surveys. Tielker et al. (2021) utilized Unipark for a national online survey in Germany [12].

Qualtrics is a cloud-based survey and experience management platform founded in 2002. It is widely used for creating and distributing surveys, collecting data, and analyzing results across various domains. Qualtrics primarily functions as an online survey platform, but it also offers an offline survey feature through the Qualtrics Offline Surveys apps.

3.2   Comparative Analysis of KoboToolbox and Unipark/Qualtrics and Paper-Based Platform

In child maltreatment studies, digital platforms provide several advantages over paper-based methods. In general, digital platforms offer a superior alternative to paper-based surveys by reducing costs, saving time, and providing real-time results. They eliminate printing and data entry expenses while enabling instant survey deployment and immediate analysis, which significantly speeds up decision-making (Table 1).

Table 1. Comparison between KoboToolbox and Unipark/Qualtrics and paper-based survey

	KoboToolbox	Unipark	Qualtrics	Paper-based survey
Platform	Open-Source Platform: open-source data collection tool, it is freely available for use and can be customized to suit specific needs	Commercial Software: Unipark/Qualtrics is a commercial survey software, and access to its features typically requires a subscription or license		Paper-based survey is a traditional way to collect data in public health and social sciences
Survey design	KoboToolbox offers a user-friendly web interface for survey design, enabling users to create various question types, skip logic, and data validation	Unipark/Qualtrics offers a comprehensive survey design interface with advanced question types, skip logic, and branching capabilities e.g. time spent for each question or tool to enhance data cleaning		The researcher designed the questionnaire on Microsoft Word then printed the questionnaire out, and photocopied documents to provide a largenumber of participants
Data collection	Supports multiple distribution methods, is collected in real-time, and data is stored securely	Unipark provides various distribution channels for survey deployments, such as email invitations, QR codes, and social media but not for mobile apps	Supports multiple distribution methods, is collected in real-time, and data is stored securely	The researcher should ask respondents or respondents to fill questionnaire on paper Requiring extra support from researchers to ensure smooth progress Time consuming, researchers must entry data after data collection

(continued)

Table 1. (*continued*)

	KoboToolbox	Unipark	Qualtrics	Paper-based survey
Apps and language on apps	App or link for entry data available English, Vietnamese, and other languages are available	Need web link, no mobile apps German, Dutch, and English are available. Vietnamese can be applied but needs to transfer to rich text format and sometimes error	App or link for entry data available English, Vietnamese, and other languages are available	Not applicable
Required internet during data collection and devices	The platform supports offline data collection Need tablet/computer use on Android devices	Internet is mandatory stable and available during data collection Need tablet/computer use Android devices	Offline **data collection** is made possible by tools; Need tablet/computer use both IOS and Android devices	No relation to the Internet No need for any devices, easy for study

(*continued*)

These rest three types of platforms are comparable in function and available features (for example, handling extensive surveys, sharing data, instantly updating edits, etc.). In general, Unipark/Qualtrics is better than KoboToolbox regarding design, data security, and paper-based survey; it offers advanced data testing and checking tools. However, the KoboToolbox aids the data cleaning and analysis by allowing easy information migration between platforms.

Digital tools like KoboToolbox, Unipark, and other DSTs are very convenient for reviewing and editing questions, collecting and organizing data, and transferring that data to other software for analysis. We can also work on multiple different datasets at once with ease; this is hugely helpful as we had conducted a pilot before our actual study, and may want to view its data alongside those of more recent runs. Other features on these platforms like real-time updates and translation help make teamwork and collaboration seamless.

Other platforms such as Unipark or Qualtrics share the latter's convenience and ease of operation. This is helpful both for us as researchers and for our study participants, who are young and may not want or be able to navigate a complicated interface. It also lends itself well to collaboration with its project project-sharing and real-time feedback features. One strength Unipark and Qualtrics have in data collection is its capability to perform automated testing and plausibility checks, which aids in producing high-value qualitative data; this is especially valuable for a complex study like ours. All copyrighted

Table 1. (*continued*)

	KoboToolbox	Unipark	Qualtrics	Paper-based survey
Data security and analysis	KoboToolbox emphasizes data security and privacy Kobotoolbox offers simple real-time cross-tabulation	Better than Kobotoolbox in data security, making it suitable for handling sensitive information Unipark/Qualtrics offers robust data cleaning and analysis tools, including filtering, cross-tabulation, and statistical analysis		Low data security and high risk of wrong answer/ skipping questions Need more time to clean the data [on paper], no real-time data available Need time/human resources to entry data and high risk of error in data analysis
Data visualization	KoboToolbox provides basic data visualization options, but it may require external tools for more in-depth data analysis and visualization	Unipark/Qualtrics provides customizable survey reports, charts, and graphs, allowing users to present their findings effectively		No data visualization at the the of data collection Need other support analysis tools to visualize data

platforms can offer better data security and are locked behind a paywall. It also requires an internet connection to function. These requirements can become barriers to conducting research in a developing country like Vietnam. Small and not well-funded studies may need help to afford Unipark.

In Vietnam, the internet connection may only be available in some areas, leading to difficulties in creating and revising an online questionnaire and using it to collect data. For our study, which relies on a lengthy, complex questionnaire as our only survey tool, Unipark's dependence on the internet can become a huge hindrance. It can even limit the diversity of the data we can collect.

Specifically for sensitive studies on child maltreatment, the KoboToolbox is more accessible and friendlier for a young study population in rural Vietnam. On the other hand, while not as secure as commercial platforms, KoboToolbox provided adequate security and was cost-effective for the study's needs. The integration of spatial data also helped identify geographic patterns of child maltreatment, informing future interventions.

As for our population of interest, which is children 12–15, these tools' displays are easy to understand and navigate; not only can this lessen errors made while answering questions, but it can also motivate the children to complete the lengthy questionnaire, as it makes the process seem more straightforward, more manageable and more enjoyable. Additionally, the fact that the questionnaire is digital (as opposed to being on paper) means it is a lot faster and more feasible to deliver it to children all over the country, thereby making our data more inclusive, diverse, and representative of Vietnam's population.

KoboToolbox is a free platform and it can work both online and offline. Regarding our study population, having a low-cost tool that can function online and offline is hugely convenient. This allows us to collect data regardless of an area's internet connection; this is critical in a developing country like Vietnam, where access to the internet is still a luxury in many places. Another way KoboToolbox helps enrich our data is via its ability to provide spatial context. This can help us identify patterns in child maltreatment throughout the country and recognize any influence location may have over said patterns. Moreover, this spatial data can help future studies and intervention programs know which areas to focus on first.

On the other hand, one disadvantage of using KoboToolbox is that to access the survey, some electronic devices using the Android operating system must be used in the schools. However, such devices are only available to some, especially in a developing country like Vietnam. Another disadvantage is that although KoboToolbox offers adequate security and is cost-effective for the study's needs, it has no copyright, which means our data is more vulnerable to corruption or theft. The following section will outline the researchers' experiences with using digital survey tools (DSTs) and how we addressed the limitations encountered.

3.3 Researcher's Experiences in Three Phases of the Study

The process to decide data collection platform for our study was very extensive. We shared our experience of deciding which DSTs for data collection in three phases of the child maltreatment project.

Preparation phase: The team consisting of members from both Vietnam and Switzerland, designed the same questionnaire for data collection and discussed which platform to use for this sensitive project. Both teams agreed to conduct data collection using tablets in schools to ensure participant privacy and safety. The Swiss team decided to use the Unipark platform because they held the Unipark license, and also shared the usage rights with the Vietnam team.

Before the commencement of a field-based pilot study involving students, a methodical and phased protocol was rigorously adhered to. Initial stages encompassed the evaluation of the survey instrument through engagement with researchers, thereby affording iterative enhancements through the integration of their valuable feedback. Subsequently, a controlled subset of 15 students was involved, allowing for the calibration of procedural intricacies.

In Vietnam, after testing with our team and 15 students in Hanoi using a paper-based method, we decided to switch to a digital platform due to the sensitive topic and errors in the questionnaire's skip logic. The Vietnamese researchers designed the questionnaire on both Unipark and KoboToolbox. After testing a small sample of students with both platforms in Hanoi, we compared their performance. While the interfaces of the two platforms are relatively similar, the question selection tablet's window in Unipark can only display 2 to 3 questions at a time. In contrast, the KoboToolbox allows for customizable lengths. The abundance of windows in Unipark led the small team during testing to find it more time-consuming due to continuous 'next' button clicks, while the KoboToolbox interface allows for seamless scrolling. Furthermore, Kobotoolbox is more language-friendly for Vietnamese than Unipark because we have to transfer the Vietnamese text to Rich-text format while Kobotoolbox can set the questionnaire to the platform directly. Additionally, concerns about the necessity for constant internet connectivity when using Unipark posed an issue for local data collection. For example, the assistant researcher swiftly addressed internet fluctuations, ensuring uninterrupted engagement when collecting data online.

Therefore, the decision to offline data collection was made. Finally, we ultimately opted for KoboToolbox for the pilot in the field site in Bac Giang province.

The second phase- a pilot survey in one secondary school was orchestrated meticulously, representing a critical facet of our research enterprise. To conduct the survey, at least 45 electronic devices using the Android operating system must be used in schools. Therefore, we would need to explore options such as providing or renting the necessary tablet devices using the Android system and establishing a private room for students to fill out the questionnaire, addressing the limitation of inadequate access to devices in rural schools in Vietnam. Finally, in this phase, we decided to rent 55 Android tablets for data collection.

This preliminary investigation comprised a cohort of 241 students, effectively priming the environment for subsequent undertakings. Notably, each iteration facilitated interactions with 40–45 participants, leading to a cumulative total of six class-based data collection rounds during this preliminary phase.

As the predetermined juncture of implementation arrived, the academic expanse was artfully transformed into an arena of intellectual exploration. An array of 45 tablets, strategically positioned, served as the conduit to channel participants into the realm of inquiry. Facilitating this transition was notably executed through integrating the KoboToolbox application, a pivotal nexus to our intricate questionnaire. An intricate voyage unfurled, spanning five discrete sections and entailing 67 interconnected pages. This digital odyssey, facilitated within the user interface of the tablets, seamlessly synthesized technology and pedagogy.

In the third phase - the main survey at 8 secondary schools:
In the main study, the number of adjusted questions reached 996 items/lines [equivalent to 98 printed pages with 338 skip items]. Due to the large sample size, data collection was conducted simultaneously in two classes, and using DSTs was much more convenient compared to paper-based research. The researchers rented 100 tablets and arranged about 40 students/room/time and two rooms were arranged at the same time to speed up the data collection (20 backup tablets for charging and device replacement purposes). However, the journey encountered challenges, underscored by a committed team of three to four assistant researchers who guided participants through complexities and maintained seamless digital progress. Notably, they had to check the battery

of the tablets carefully, assist students who wanted to revise their answers in the previous tablet window, and handle the skip logic questions [already recorded] if there were changes in the responses. The researchers had to ensure that after completing the survey, adolescents clicked submit, confirmed that their responses were recorded, and school students were guided to leave the room quietly to avoid disturbing other students still completing the survey. On average, student respondents completed the questionnaire for 35–65 min. This temporal variance was contingent upon the number of child maltreatment they encountered, thus signifying the evolving nature of participant engagement. The dynamic nature of this enterprise was further fortified through the incorporation of real-time data updates, thereby granting the data manager the capability to monitor the progression and the count of completed surveys. After finishing data collection, researchers confirmed all the records were uploaded to the cloud and all the Kobocollect apps were removed before returning to the renting agencies.

4   Discussion

KoboToolbox has strengths and limitations (2,4,5). Within studies concerning child maltreatment, technological aspects may become a barrier to both tools' application. Some children may struggle to gain access to surveys on these digital tools due to a lack of access to electronic devices or if they are facing maltreatment and are denied access to these tools. Due to the young age of selected participants, some may have a more challenging time following instructions, staying focused, or even understanding what they are being asked to do by the researchers. When using these tools in Vietnam, researchers must consider their accessibility.

In our specific research context, the KoboToolbox is a preferable option for researchers in Vietnam. Given the substantial questionnaire expected to yield significant data, researchers need a platform solution that supports online and offline operations, a capability uniquely provided by KoboToolbox.

Using KoboToolbox has greatly enhanced our research by ensuring transparent adherence to ethical procedures and providing a secure environment for students to respond to questions about child maltreatment. Although KoboToolbox's data security measures may not be as robust as those of Unipark or Qualtrics platforms, its cost-free availability allows us to conduct this comprehensive nationwide study with a lower financial burden. The range of data-centric features offered by KoboToolbox meets the needs of our study effectively.

Given the extensive questionnaire, online data collection has proven invaluable, preventing students from facing discrepancies during question transitions and improving the accuracy of their responses and disclosures. The design's anonymity has encouraged students to share information about maltreatment candidly. Students have found the software user-friendly and remarkably resilient, even without internet connectivity, as data updates continue to function offline. Notably, the continuous data updating and visual representation of raw results through tables and charts provide immediate feedback in the field, allowing stakeholders to gain preliminary insights that a traditional paper-based study would not facilitate.

In general, the DSTs are becoming an indispensable part of data collection and management in our rapidly digitalizing world. They help make these processes more streamlined and less prone to errors (compared to if these surveys were conducted on paper and the data transferred to storage software later). DSTs such as Kobotoolbox can efficiently run complex questionnaires and gather lengthy data sets, which is beneficial when studying a population as extensive and encompassing as children. As research in Vietnam becomes more advanced (in terms of funding, organizing, *etc.*), implementing these tools will become more accessible, if not necessary, to carry out more complex and wider-scale projects.

Some limitations should be addressed. The tool's data security features, while sufficient, are not as robust as those of commercial platforms like Unipark or Qualtrics, which may be a concern for handling sensitive information such as child maltreatment data. Additionally, some challenges were noted with internet fluctuations during data collection, requiring extra support from researchers to ensure smooth progress and the security of data in the free platform.

In the future, multilingual tools can open up more opportunities for international research, acting as bridges between Vietnam and other nations, allowing all to exchange information and collaborate on projects much faster and more smoothly. Particularly for Vietnam as a developing country, we can take advantage of this improvement in communication to enrich our research; one example is how, through KoboToolbox, we could reference the Swiss partner's questionnaire on child maltreatment while working on our version. Another helpful feature of the KoboToolbox is to provide spatial context to the data, which allows our study results to inform future interventions and solutions for child maltreatment and identify areas having a prevalence of child maltreatment. Researchers distribute countermeasures more equitably (for example, maltreated children in poorer, more rural areas can begin to receive the help and support that those in more affluent, more urban areas have been able to access due to the latter's advantages in wealth, distance, and proximity to where child maltreatment studies are being conducted).

5   Conclusion

In conclusion, the selection and utilization of KoboToolbox for the collection of child maltreatment data in Vietnam were driven by several compelling factors. The platform's versatile interface, which accommodates longer question sequences and offers the convenience of screen-scrolling, emerged as a significant advantage over paper-based surveys and other DST platforms. The streamlined user experience on KoboToolbox, enabled by its user-friendly navigation, proved to be a time-efficient solution during testing – a vital consideration for research endeavors of this nature. Furthermore, the offline data collection feature of KoboToolbox addressed the critical challenge of consistent internet connectivity often encountered in local contexts, overcoming a potential hurdle faced by other digital platforms.

Acknowledgment. This research is funded by the Vietnam National Foundation for Science and Technology Development (NAFOSTED) and Swiss National Science Foundation (SNFS) under grant number IZVSZ1.203300.

Disclosure of Interests.   The authors declare that they have no competing interests.

References

1. Gulde, M., Köhler-Dauner, F., Mayer, I., et al.: Negative effects of the SARS-CoV-2 pandemic: the interlinking of maternal attachment representation, coping strategies, parental behavior, and the child's mental health. Front. Pediatr. **10**, 939538 (2022)
2. Adhikari, B., Poudel, L., Thapa, T.B., et al.: Prevalence and factors associated with depression, anxiety, and stress symptoms among home isolated COVID-19 patients in Western Nepal. Dial. Health. **2**, 100090 (2023)
3. Unipark. Shop for buying or test a license. Accessed 30 Aug 2023. https://www.unipark.com/umfragesoftware-bestellen/
4. Meinck, F., Steinert, J., Sethi, D., et al.: Measuring and monitoring national prevalence of child maltreatment: a practical handbook. In: World Health Organization Regional Office for Europe, Denmark, p. 49 (2016)
5. Vasantha Raju, N., Harinarayana, N. (eds.): Online survey tools: A case study of Google Forms. National conference on scientific, computational & information research trends in engineering, GSSS-IETW, Mysore (2016)
6. KoboToolbox. Features of KoboToolbox. Accessed 15 July 2023. https://www.kobotoolbox.org/features/
7. Unipark. Top 10 Features Unipark. Accessed 18 July 2023. https://www.unipark.com/en/survey-software/
8. Uhlig, C.E., Seitz, B., Eter, N., et al.: Efficiencies of Internet-based digital and paper-based scientific surveys and the estimated costs and time for different-sized cohorts. PLoS ONE **9**(10), e108441 (2014)
9. Rose, M., Devine, J.: Assessment of patient-reported symptoms of anxiety. Dial. Clin. Neurosci. (2022)
10. Thriemer, K., Ley, B., Ame, S.M., et al.: Replacing paper data collection forms with electronic data entry in the field: findings from a study of community-acquired bloodstream infections in Pemba, Zanzibar. BMC Res. Notes **5**(1), 1–7 (2012)
11. Lätsch, D., Thi, L.M., Huong, N.T.: The role of schools in detecting and responding to child maltreatment: a collaborative approach to evidence-based policy development in Switzerland and Vietnam. Accessed 20 July 2023. https://data.snf.ch/grants/grant/203300
12. Tielker, J.M., Weber, J.P., Simon, S.T., et al.: Experiences, challenges and perspectives for ensuring end-of-life patient care: a national online survey with general practitioners in Germany. **16**(7), e0254056 (2021)

An Approach for Building Question-Answering Systems Using Large Language Models on Medical Domain Dataset

Hong-Viet Tran[1]([✉]), Lam-Quan Tran[2]([✉]), Minh-Hoang Tran[1],
Van-Thuy Mai[2], and Van-Tan Bui[3]

[1] Institute for Artificial Intelligence, University of Engineering and Technology,
Vietnam National University, Hanoi, Vietnam
`thviet@vnu.edu.vn`
[2] Digital Health Center, Hanoi University of Public Health, Hanoi, Vietnam
`{tlq,mvt}@huph.edu.vn`
[3] Faculty of Information Technology, University of Economics - Technology for
Industries, Hanoi, Vietnam
`bvtan@uneti.edu.vn`

Abstract. This study introduces an approach for constructing a specialized question-answering system for the medical domain through the application of large language models. To improve answer accuracy and reliability, the system adopts the retrieval-augmented generation framework, which enables the integration of external knowledge sources by retrieving relevant information to support response generation. Combined with prompt design and reinforcement learning techniques, the model adapts to updates in medical data. Using a large Vietnamese language model, the system achieves high performance in diagnosing diseases and answering medical-related queries. Experimental results show that the system provides accurate and context-aware answers in the field of healthcare consultation, demonstrating its potential to improve medical chatbot systems. Key techniques include the integration of a multi-agent architecture, long-term memory, and a hybrid search method for better information retrieval, ensuring personalized and reliable responses for users.

Keywords: LLM-Large language models · chatbot · RAG-Retrieval Augmented Generation · AI-Artificial Intelligence · NLP-Natural Language Processing

1 Introduction

Currently, when Meta AI introduced the approach of combining chatbots with large language models[1], it can be said that most chatbots were in the form of

---

[1] https://llama.meta.com/code-llama.

H.-V. Tran and L.-Q. Tran—Equal contribution.

© ICST Institute for Computer Sciences, Social Informatics and Telecommunications Engineering 2026
Published by Springer Nature Switzerland AG 2026. All Rights Reserved
S. Thomassey et al. (Eds.): RAIDS 2024, LNICST 673, pp. 48–58, 2026.
https://doi.org/10.1007/978-3-032-14055-5_4

live chat sales, product introductions, and follow-up customer chats good script according to keywords. For chats that do not have keywords or are not scripted, chatbots often, on the one hand, cannot provide answers and, on the other hand, cannot interact naturally with humans. This challenge comes from many factors, including understanding conversation, in other words, understanding language and linguistic context. Language is basically a complex system, including grammatical rules and different ways of expressing each entity. When applying NLP, a subfield of artificial intelligence and machine learning, to textual data, chatbots gain the ability to comprehend user input, understand contextual meaning, and generate human-like responses [1]. At the core of most NLP systems lies the language model, which is typically a probabilistic model trained on natural language input with the objective of predicting the next word or filling in missing elements within a sentence [2]. As the number of model parameters increases, reaching tens or even hundreds of billions—these models demonstrate not only improved syntactic and semantic understanding but also enhanced capabilities in contextual reasoning and natural language generation. To distinguish models with a massive number of parameters, the term large language models has been adopted by the research community[2]. Representative examples of LLMs include commercial platforms such as ChatGPT, BingAI, and Google Bard, as well as open-source counterparts like LLaMA [3], Alpaca [4], and Vicuna [5]. These models are trained on extensive text corpora, enabling them to effectively capture complex semantic representations and learn advanced syntactic and contextual patterns. Although the large language model approach to the question-and-answering problem shows applicability across many languages, there is still a gap for Vietnamese, a language with its own syntactic and contextual characteristics. Furthermore, our chatbot application depends on the domain: the field of health diagnosis and consultation, related issues in the medical field (Fig. 1).

In addition, a disadvantage of large language models is that they cannot answer updated data as well as new data. When encountering these questions, chatbots often create incorrect answers. Therefore, in this article, we apply the RAG method to limit incorrect answers, combined with prompt design and Reinforcement Learning techniques every time the data are updated, to answer questions. Answers to medical industry questions and new data.

The main contributions of the study are:

- Building a question-answering system to automatically answer questions related to QA-medical dataset about health consultation and disease diagnosis and related issues in the medical field.
- Using a large Vietnamese language model system and applying RAG architecture to reduce the incorrect answers. Additionally, we also use some technical to enhance RAG queries (Fig. 2).

---

[2] https://www.techfund.one/p/nvidia-amd-and-the-ai-cycle.

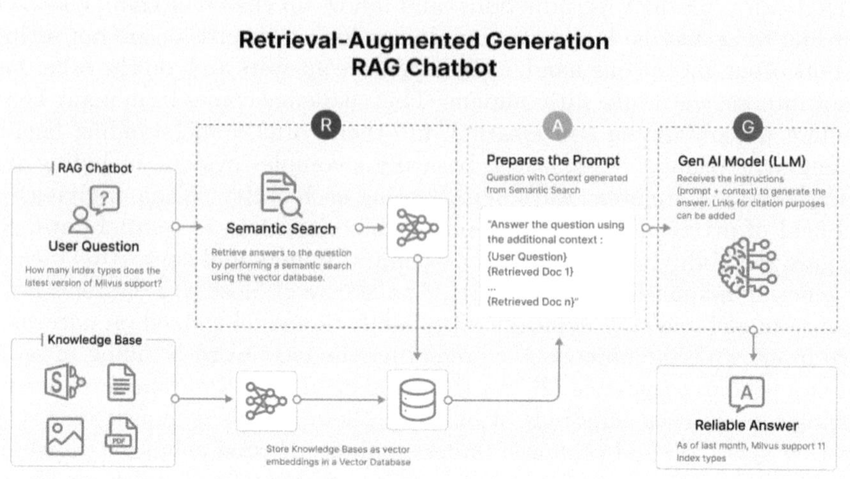

Fig. 1. Architecture of the LLM-based Q&A system, integrated with an embedding module for querying and extracting relevant input information to generate factually grounded and reference-supported responses. **Source:** https://zilliz.com/learn/Retrieval-Augmented-Generation

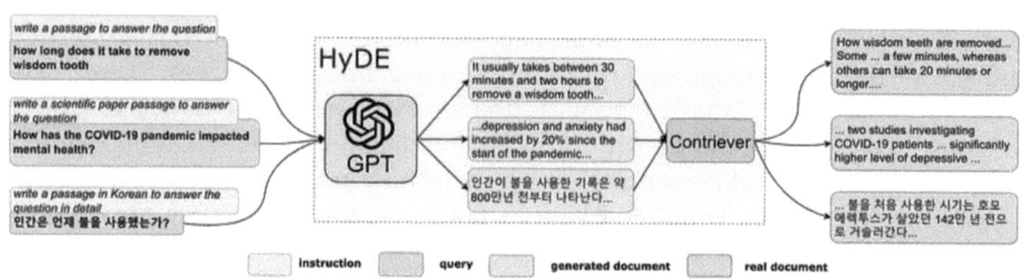

Fig. 2. An illustration of the HyDE model. **Source:** *Precise Zero-Shot Dense Retrieval without Relevance Label* [6].

2   Related Work

The advancement of large language models has garnered considerable interest within the artificial intelligence research community, prompting extensive investigations into their architecture, training strategies, and cross-linguistic adaptability across a variety of languages [7,8].

Recent studies have investigated diverse methods and approaches to perfect large language models, standardize training data sets, and improve performance in specific tasks and domains [9–11].

Research on Vietnamese language processing has been carried out preprocessing and training Vietnamese monolingual language models [12], with applications for QA Question Answering tasks [13]; Applications on NER, named entity recognition [14]; Applications that automatically summarize text [15], ...

The Retrieval-Augmented Generation paradigm integrates generative Large Language Models with conventional information retrieval mechanisms. This synergy allows LLMs to produce responses, explanations, and instructions that are contextually relevant and grounded in external knowledge sources. A typical RAG configuration comprises three main components: a vector-based search engine (e.g., ElasticSearch), an embedding model for semantic representation, and a generative LLM for response synthesis. Although the overall architecture is relatively straightforward to implement, its real-world performance is highly dependent on the effectiveness and accuracy of the retrieval module (Fig. 3).

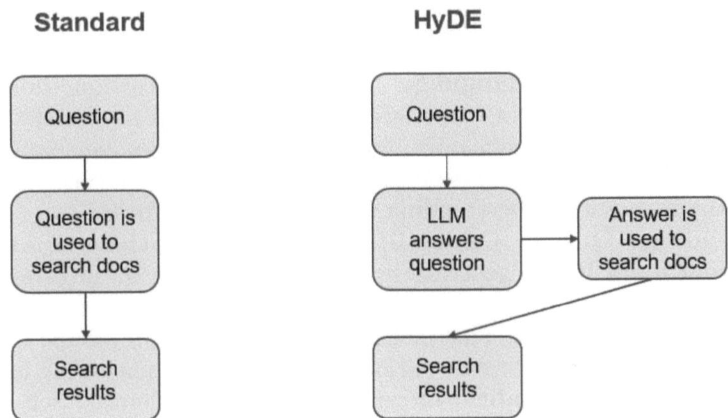

Fig. 3. Comparison between the standard retrieval process and the hypothetical document embeddings approach.

The RAG framework is structured around two core components: the retriever, which locates and extracts relevant documents from a vector-based database, and the generator, typically a large language model, which composes the final response using the retrieved content. The Hypothetical Document Embeddings (HyDE) approach extends and enhances the standard RAG pipeline through several notable improvements:

- **Generating Hypothetical Documents**: HyDE generates a hypothetical document based on the input query, which serves as a proxy to retrieve relevant documents. This technique helps encapsulate the semantic intent of the query more effectively than directly relying on initial retrieval results.
- **Handling Ambiguous Questions**: For queries that are vague or contextually unclear, HyDE enriches the query by adding additional context using LLM capabilities, thereby improving the quality of subsequent retrieval and generation.

- **Optimizing Document Queries**: Given that most document repositories contain answers rather than questions, HyDE leverages a hypothetical answer as a retrieval query. This shift enhances retrieval precision and relevance.

HyDE plays a significant role in NLP tasks by enabling relevant document retrieval without requiring supervised training or labeled data. Its capacity to generate contextually rich hypothetical content makes it particularly effective in applications such as multilingual web search and question answering.

To the best of our knowledge, at the time of this writing, this work represents one of the earliest efforts to integrate a Large Language Model with a chatbot system using the RAG methodology within the Vietnamese healthcare domain, and to demonstrate its applicability in real-world scenarios.

3 Methodology

3.1 Methods Technical Specifications

- **Integrating large language models:** A medical question-answering dataset is utilized to fine-tune large language models on the QA generation task. This process significantly improves the chatbot's capacity to comprehend and generate natural language responses, while simultaneously enriching its domain-specific understanding in the field of healthcare.
- **Extending Query Capabilities with Hypothetical Document Embeddings:** HyDE generate hypothetical answers by utilizing large language models to create responses based on related text or available information from the question. This method potentially allows for the expansion of search capabilities to explore various possibilities for answering questions.
- **Improve answer quality:** Integrate methods like Rerank, Self-consistency with Chain-of-Thought (CoT), ReAct, Sentence Window Retrieval, and Auto-Merging Retrieval to enhance the performance of a Q&A system. This approach enables the chatbot to deliver precise, comprehensive, and contextually relevant responses.
- **Applying Multi-Agent architecture:** To support diverse consulting services and training industries, we apply Multi-Agent architecture for chatbot. Each agent will specialize in a specific field, helping Q&A system to generate accurate and in-depth answers.
- **Integrated Long-term Memory:** The Q&A system is integrated with the Long-term Memory feature to remember and personalize knowledge for each user. This helps the Q&A system maintain a seamless conversation and provide information tailored to each individual's needs.
- **Supports multi-modal interaction:** The Q&A system supports conversion for text and voice in term of both input and output through Text-to-Speech and Speech-to-Text features. This helps users interact with chatbots conveniently and flexibly.
- **Ensuring safety and transparency:** Content Moderation is applied to ensure that chatbot output content is safe and appropriate by using moderation APIs or filtering sensitive words based on a sentitve word list. At

the same time, the system also has the ability to cite and acknowledge document sources (Citations and Attributions) to ensure the transparency and reliability of the provided information.

- **Search optimization:** we used Hybrid Search method, which combines full-text search and vector search, then rerank to select the results that best match the user's query. This helps chatbots find information quickly and accurately.
- **Multilingual processing:** Using a large language model allows the chatbot to process documents in multiple languages without translating them into Vietnamese. This helps save time and effort, while expanding accessibility beyond the Vietnamese language domain.
- **Solving the problem of new and updated data:** To solve the problem of large language models not responding correctly to new and updated data, we apply the RAG method, combined with Prompt Design technique so that the data is update in anytime. This helps the question and answer system to adapt better to new and specific information in the medical field.

3.2 Architectural Design

Currently, the limitation of embedding models for Vietnamese language when the context length is the short context length of 512 tokens", leading to low performance when chunking the document by conventional methods. To solve this problem, we proceed to split documents based on titles to cover as much data as possible. The split data with large context length is stored as metadata. In the process of collecting essential information and achieving high results, we perform summarizing the document, thereby reducing the size of the original document while maintaining the overall meaning of the document. At the same time, questions are generated based on these documents. The summarized data and generated questions are used for semantic search (Fig. 4).

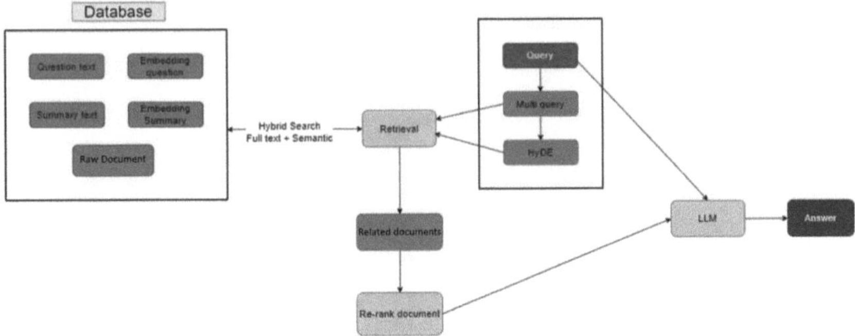

Fig. 4. Pipeline for process with the Hypothetical Document Embeddings.

In fact, the same idea can be achieved in many different ways. With different questions, the returned results will be different. To solve this problem, we use query expansion techniques. Particularly, we paraphrase the initial user query into many versions then search the related documents on these paraphrased queries, helping us to find more diverse documents. With the classic RAG method, we will directly use the user's questions to search for similarities in the database. However, to improve, we use the HyDE method. This method creates hypothetical answers by using large language models to generate answers based on related text or on information available from the question. These questions may or may not be correct, but allow predictions about possible answers that are allowed to expand the search range of different possibilities to answer the question. The retrieval process is executed based on the combination of keyword-based search and semantic search on each pair of this. Keyword search uses the BM25 algorithm to find relevant documents and semantic search is performed by the bi-encoder model comparing the similarity of embedding pairs. The unified search process, using a combination of the two above search methods with HyDE process brings optimal results. We conduct a combined search with the query searched based on the previously generated questions of HyDE with the document summary. The final result of the document retrieval process the average of the search methods. Then, we proceed to remove duplicates to obtain a list of the most relevant documents to the query.

The retrieval process is optimized for searching for existing relevant documents stored in the database. However, this also limits us from finding truly meaningful documents because the retrieved documents have quite a lot of documents which are similar to each other in the query results. This does not make sense when these documents are fed into a large language model. Instead, we will select the best documents to include in the LLM context. To do this, we use the Maximal Marginal Relevance (MMR) method to re-rank the retrieved documents and select the top documents for the LLM context. Re-ranking is an iterative process with each step performing the following steps: - Compute the similarity of the documents that we need to re-rank D_i to the query vector Q.

$$Similarity\ to\ Query = sim(D_i, Q) \tag{1}$$

- Compute the similarity of the vectors that we need to re-rank D_i to the re-ranked vectors D_j

$$Similarity\ to\ ranked\ vectors = sim(D_i, D_j) \tag{2}$$

- We will take the vectors that maximize the combination of both measures.

$$MMR = arg \max_{D_i \in R \setminus S} [\lambda sim(D_i, Q) - (1 - \lambda) \max_{D_j \in S} sim(D_i, D_j)] \tag{3}$$

The purpose of the re-rank process is to help us find vectors that correspond to the query but are different from the re-ranked vectors. By repeating this process many times, we will re-rank the vectors that corresponds to the best documents. After performing the re-ordering of the retrieved documents, we get the top k documents that are most relevant to the combined questions. Then, the user's query combined with the retrieved documents are fed into the large language model that gives the final answer to the user.

4   Training Dataset

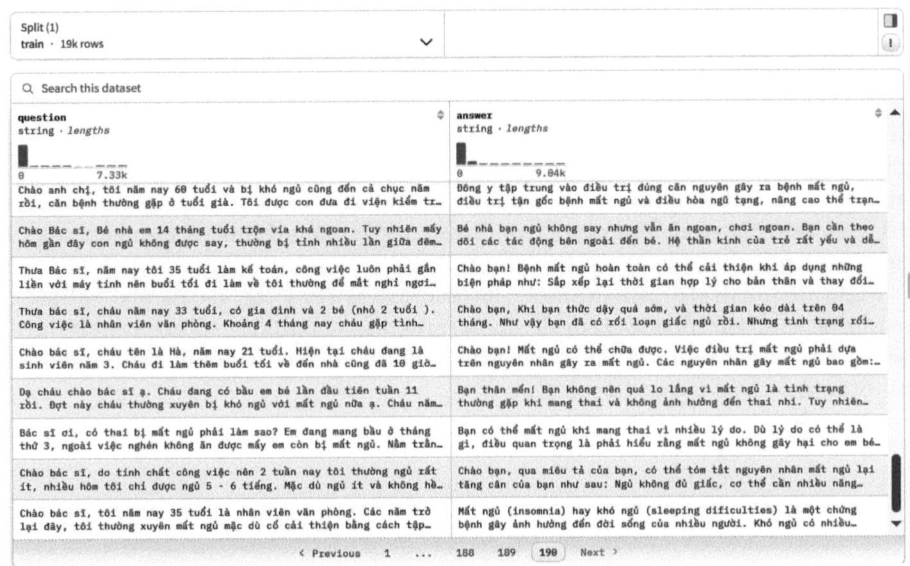

Fig. 5. An example of Q & A data in the fine-tuning dataset.

Training and Testing Dataset: To construct the training and evaluation datasets, the research team leveraged an open-source Vietnamese medical question-answering dataset, as introduced earlier. For fine-tuning, the team employed the Vistral-7B-Chat model [16], a multi-turn conversational large language model tailored for the Vietnamese language. This model is an extension of Mistral-7B [17], which has been further trained using diverse data sources via continual pre-training and instruction tuning. Notably, Mistral-7B-v0.1, a generative model with 7 billion parameters, outperforms LLaMA 2 13B across multiple benchmarks.

The primary dataset used is the Vietnamese Healthcare Question Answering dataset[3], comprising 9,334 QA pairs. These question-answer pairs were collected from users seeking health-related information on reputable websites, with responses provided by certified medical professionals. Specifically, data was sourced from https://edoctor.io/hoi-dap and https://www.vinmec.com/vi/tin-tuc/hoi-dap-bac-si/. After merging and cleaning data from both platforms, the final dataset included 9,335 QA pairs, as illustrated in Fig. 5.

To enrich this dataset with deeper domain-specific knowledge, particularly in disease diagnosis and treatment, the research team adopted the approach presented in [6]. The augmented dataset was subsequently partitioned into training

---

[3] https://huggingface.co/datasets/thviet79/dataset_medical.

and test sets to serve the fine-tuning and evaluation phases. These datasets were used to develop and assess a chatbot system powered by large language models, focused on supporting diagnostic and treatment-related queries in the public healthcare domain.

5   Experimental Results

The experiment was carried out by integrating a large language model, specifically the Vistral-7B-Chat, which was fine-tuned on a medical-domain Q&A dataset. This fine-tuning process enhances the chatbot's ability to comprehend and generate natural language responses, while simultaneously expanding its domain-specific knowledge in healthcare.

Figure 6 illustrates an example interaction within the developed question-answering system. The system demonstrates the capability to accurately interpret and respond to questions concerning health consultation, disease diagnosis, and other medical-related topics.

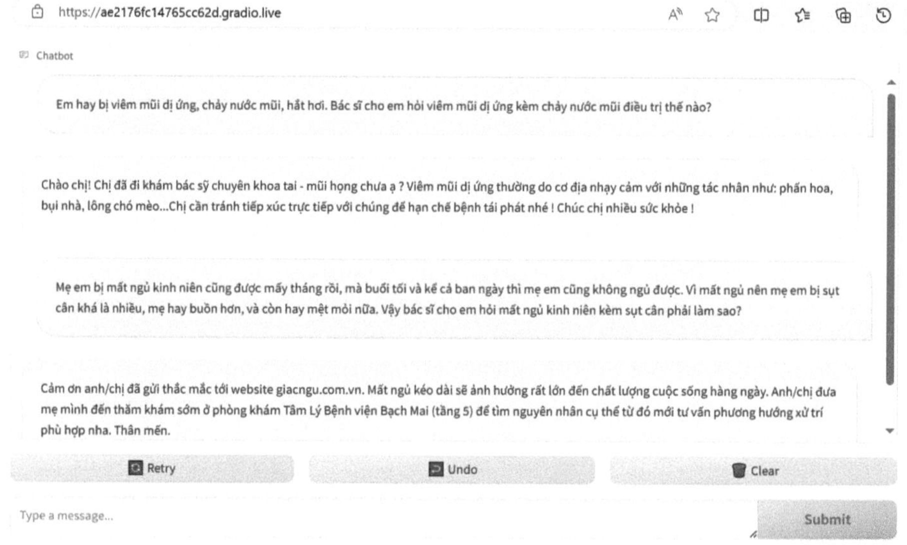

Fig. 6. Example of the question-answering system utilizing a Vietnamese medical dataset.

To assess the performance of the automatic answer generation model, we employed ROUGE (Recall-Oriented Understudy for Gisting Evaluation), a commonly used evaluation metric for natural language generation tasks, including question answering. ROUGE measures the overlap between system-generated summaries and human-written reference summaries, emphasizing the extraction of semantically important content from the source text.

The model's performance, measured using ROUGE scores, is presented in Table 1.

Table 1. Model Results with ROUGE Scores

	ROUGE-1	ROUGE-2	ROUGE-L
Precision	0.65	0.48	0.51
Recall	0.76	0.59	0.61
Fmeasure	0.61	0.46	0.48

6 Conclusion

In this article, we proposed an approach to building a question and answer system using large language models on medical domain data with question and answer techniques using the RAG method to limit answers inaccurate, combined with prompt design and Reinforcement learning techniques every time the data is updated, to answer questions in the medical field. Experimental results show that the question and answer system using the large Vietnamese language model achieves high efficiency in making diagnoses with medical domain data. The question and answer system has the ability to understand and accurately answer questions related to health advice, disease diagnosis and issues related to the medical field.

Feedback collected from end users shows that this is a potential research and application direction in diagnostic questioning and providing advice on health and issues related to the medical field. , for example, the ability to read comprehension, recognize named entities, and the ability to interact with users shows intelligent capabilities in AI applications.

Applying techniques such as Multi-Agent, Long-term Memory and Hybrid Search helps the Q & A system provide appropriate and personalized answers for each user. However, there are still some limitations that need to be improved in the future, such as expanding the training data set, optimizing response time, and enhancing natural language processing capabilities. In addition, more research and testing is needed to evaluate the effectiveness of the Q & A system in practice.

7 Ethics Statement

The data and models used in this study do not have unethical applications or risky broader impacts. The data is widely used in the research community.

The data does not include any confidential or personal information of employees or companies. Models are accessed using the Huggingface links provided. There is no bias towards redeployment that could affect the final result.

References

1. Zhao, W.X., et al.: A survey of large language models (2023)
2. Luitel, D.: A language-model-based approach for detecting incompleteness in natural-language requirements. Ph.D. dissertation, Université d'Ottawa/University of Ottawa (2023)
3. Touvron, H., et al.: Llama: open and efficient foundation language models (2023)
4. Taori, R., et al.: Stanford alpaca: an instruction-following llama model. https://github.com/tatsu-lab/stanford_alpaca (2023)
5. Zheng, L., et al.: Judging llm-as-a-judge with mt-bench and chatbot arena (2023)
6. Gao, L., Ma, X., Lin, J., Callan, J.: Precise zero-shot dense retrieval without relevance labels (2022). https://arxiv.org/abs/2212.10496
7. Zeng, W., et al.: Pangu-α: Large-scale autoregressive pretrained Chinese language models with auto-parallel computation (2021)
8. Peng, B., Li, C., He, P., Galley, M., Gao, J.: Instruction tuning with gpt-4 (2023)
9. Koleva, A., Ringsquandl, M., Buckley, M., Hasan, R., Tresp, V.: Named entity recognition in industrial tables using tabular language models. In: Li, Y., Lazaridou, A. (eds.) Proceedings of the 2022 Conference on Empirical Methods in Natural Language Processing: Industry Track, pp. 348–356. Association for Computational Linguistics, Abu Dhabi (2022). https://aclanthology.org/2022.emnlp-industry.35
10. Qiao, L., Wu, C., Liu, Y., Peng, H., Yin, D., Ren, B.: Grafting pre-trained models for multimodal headline generation. In: Li, Y., Lazaridou, A. (eds.) Proceedings of the 2022 Conference on Empirical Methods in Natural Language Processing: Industry Track, pp. 244–253. Association for Computational Linguistics, Abu Dhabi (2022). https://aclanthology.org/2022.emnlp-industry.25
11. Chen, X., Lin, M., Schärli, N., Zhou, D.: Teaching large language models to self-debug (2023)
12. D. et al.: Vietai/gpt-j-6b-vietnamese-news (2023). https://huggingface.co/VietAI/gpt-j-6B-vietnamese-news
13. Phan, L., Tran, H., Nguyen, H., Trinh, T.H.: Vit5: pretrained text-to-text transformer for vietnamese language generation (2022)
14. Vu Xuan, S., Vu, T., Tran, S., Jiang, L.: ETNLP: a visual-aided systematic approach to select pre-trained embeddings for a downstream task. In: Mitkov, R., Angelova, G. (eds.) Proceedings of the International Conference on Recent Advances in Natural Language Processing (RANLP 2019), pp. 1285–1294. INCOMA Ltd., Varna (2019). https://aclanthology.org/R19-1147
15. Tran, C.D., Pham, N.H., Nguyen, A., Hy, T.S., Vu, T.: Videberta: a powerful pre-trained language model for Vietnamese (2023)
16. Van Nguyen, C., et al.: Vistral-7b-chat - towards a state-of-the-art large language model for Vietnamese (2023)
17. Jiang, A.Q., et al.: Mistral 7b (2023). https://arxiv.org/abs/2310.06825

Comparison of Various Models for Predicting In-Hospital Mortality Using Mimic-III Database

Pham Thao Nhu Doan[(⊠)], Linh-Chi Nguyen, and Thanh-Phuong Tran

University of Economics and Finance, Ho Chi Minh City, Viet Nam
nhudpt20@gmail.com

Abstract. Predicting in-hospital mortality is a critical problem in the medical field, aiding doctors in using appropriate treatment methods. This study focuses on evaluating and comparing various models for predicting patient mortality using the MIMIC-III database. The models examined include Logistic Regression, Random Forest, Convolutional Neural Networks (CNN), Vision Transformer (ViT), and Time-Series Representation Learning via Temporal and Contextual Contrasting (TS-TCC). The results indicate that most models achieved high accuracy, with ViT standing out due to its short training time while maintaining accuracy in comparison with other models. However, the TS-TCC model proved unsuitable in this study. The research highlights the potential for applying image processing models to time series data in medicine, opening up promising avenues for future research.

Keywords: MIMIC-III · Self-supervised Learning · Deep Learning · Healthcare · In-hospital mortality

1 Introduction

Predicting the mortality of patients during hospitalization is crucial for assessing the severity of illness and determining appropriate interventions, particularly in emergency and critical care settings. In this study, we focus on evaluating machine learning and deep learning models to predict patient mortality using the MIMIC-III database, a large and complex medical data source.

Recognizing the similarities between the time-series characteristics of patient data and those of image and conventional time-series, we propose applying different models that have been effective in these respective domains for this prediction task. These models include Logistic Regression, Random Forest, Convolutional Neural Networks (CNN)[10], Vision Transformer [2], and a novel model, Time-Series Representation Learning via Temporal and Contextual Contrasting [3]. Our objective is to identify which model performs best in predicting mortality by comparing and analyzing their outcomes.

The results indicate that most models achieve high accuracy, with noticeable differences in training times. Notably, the Vision Transformer model stands out, as it requires

S. Thomassey et al. (Eds.): RAIDS 2024, LNICST 673, pp. 59–67, 2026.
https://doi.org/10.1007/978-3-032-14055-5_5

significantly shorter training times while maintaining nearly equivalent accuracy. Additionally, we found that the TS-TCC model is not well-suited for the input data used in this study. In conclusion, the application of image processing models for intensive care units (ICU) yields highly promising and encouraging results.

2 Related Works

Mortality prediction is a common task addressed in numerous studies using various models. The GRU-D model [1] is described as a novel deep learning method designed to handle missing data in multivariate time series, particularly in the healthcare domain.

In a study by Nestor et al. [6], it was emphasized that neglecting the temporal aspect in assessments can lead to an overestimation of predictive quality. The authors demonstrated that the predictive performance of model quickly saturates with a small amount of data. Moreover, they developed new aggregate representations that are more effective in generalization.

Harutyunyan et al. [4] described the application of the Long Short-Term Memory (LSTM) model for predicting in-hospital mortality. They presented how LSTM was used to process clinical observation data and generate hidden state vectors. Additionally, they discussed the use of a channel-wise module and the rationale behind its application.

Nowroozilarki et al. [7] introduced a real-time mortality risk prediction model for the ICU using the BoXHED method on the MIMIC IV dataset. The BoXHED method non-parametrically incorporates time-dependent variables, providing an absolute measure of a patient's real-time mortality risk.

3 Methodology

3.1 Time-Series Representation Learning via Temporal and Contextual Contrasting (TS-TCC)

Eldele et al. [3] proposed using unlabeled data to learn feature representations from time series by generating transformations on the time series. These transformations are referred to as strong augmentation and weak augmentation. While weak augmentation involves multiplying the input by a random ratio to alter its scale, strong augmentation segments the time series into smaller parts, shuffles them, and adds random noise. Finally, these features are mapped to a higher-dimensional space to fit the model through a Convolutional Neural Network (CNN).

Overall, TS-TCC consists of two main modules: the temporal contrasting module and the contextual contrasting module (Fig. 1).

The temporal contrasting module learns time-consistent features through cross-view sequential prediction between the two previously mentioned 'views'. The module begins by randomly selecting a time step t, where a context vector c_t, which synthesizes information from $z_{\leq t}$, is generated using a Transformer Encoder. This vector is then used to predict future features from z_{t+1} to z_{t+K}.

Specifically, strong augmentation generates c_t^s, and the result of weak augmentation is c_t^w. Eldele et al. proposed a cross-view prediction task, using the context vector from

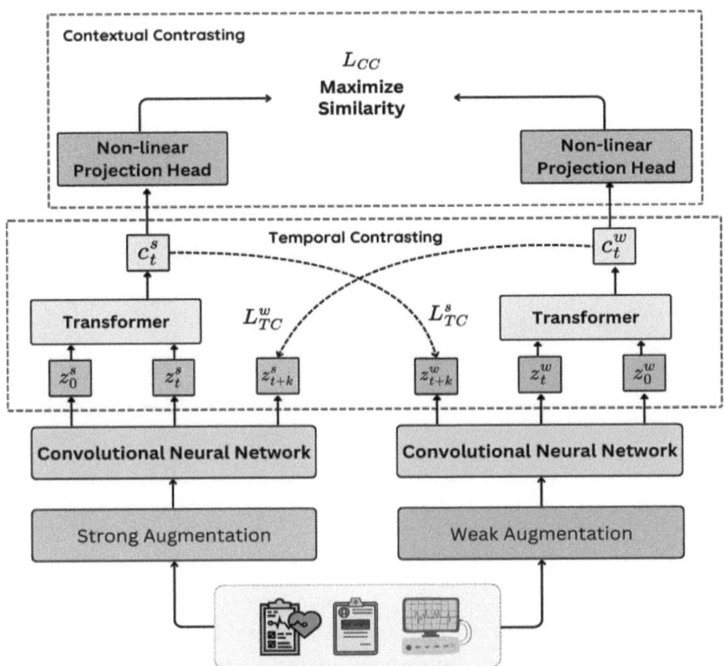

Fig. 1. Overall structure of TC-TCC model for ICU data.

strong augmentation c_t^s to predict the next K time steps for weak augmentation z_{t+k}^w and vice versa. The contrastive loss function maximizes the dot product between the prediction and the actual features of the same sample, meaning the model learns to bring the representations of predictions and actuals closer together in space. Thus, two loss functions are applied: L_{TC}^s và L_{TC}^s.

The contextual contrasting module learns distinctive features and is considered a self-supervised training model. Non-linear projection maps the context vector c_t to be compatible with this module. From the same original sample, features from strong augmentation c_t^i and weak augmentation $c_t^{i^+}$ are considered a positive pair $(c_t^i, c_t^{i^+})$. Additionally, context vectors generated from different inputs are defined as negative pairs. The contextual contrastive loss function maximizes the similarity between positive pairs. Consequently, the final result enhances the model's ability to distinguish between samples.

Thus, the total loss function for the self-supervised learning model is a combination of the temporal contrasting and contextual contrasting modules:

$$L_{unsup} = \lambda_1 \cdot \left(L_{TC}^s + L_{TC}^w\right) + \lambda_2 \cdot L_{CC}, \tag{1}$$

where λ_1 and λ_2 are constants representing the influence of the two loss functions on the total loss.

Convolutional neural network (CNN). Time series data representing the first 24 h of patient x in ICU is denoted as $x \in R^{1 \times 24 \times D}$, where D is the number of feature dimensions.

To accommodate this data size, we modified the CNN architecture mentioned earlier in the TS-TCC model from 1D to 2D and also utilized it to assess the feasibility of this network for predicting mortality (Fig. 2).

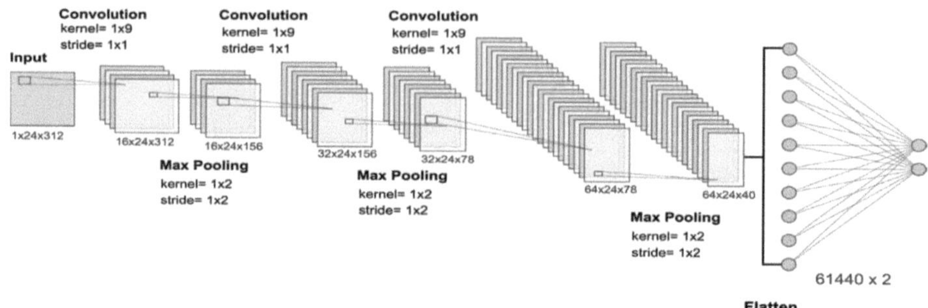

Fig. 2. Convolutional Neural Network.

The CNN employs consecutive convolutional layers to extract features from the input data, with each convolutional layer followed by batch normalization and ReLU activation functions to ensure non-linearity and stability during training. Pooling layers are applied to reduce the spatial dimensions of the features and enhance abstraction, while dropout layers are added after each convolutional block to mitigate the risk of overfitting by randomly dropping some nodes in the network.

3.2 Vision Transformer (ViT)

Dosovitskiy et al. [2] introduced the Vision Transformer (ViT) model with the aim of applying Transformer models for image classification. We leverage this model while making several adjustments to suit the objectives of our research (Fig. 3).

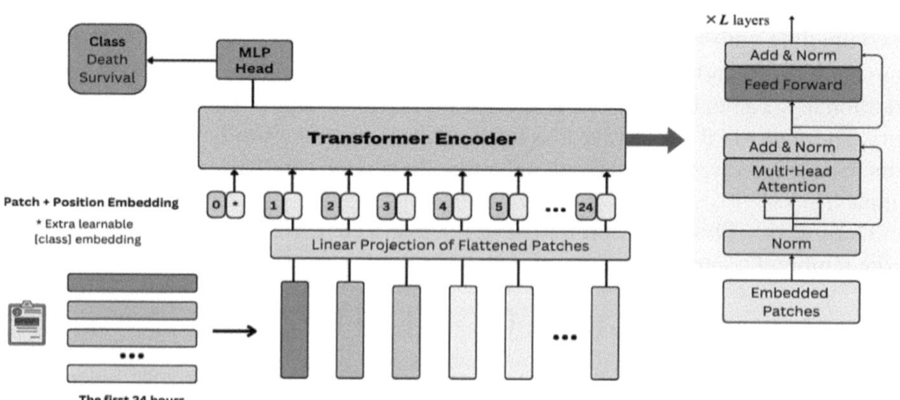

Fig. 3. ViT for Time series data.

The input data comprises a time series that captures the first 24 h of each patient's vital signs. Each hour, the patient's vital signs are recorded, forming a data sequence with the initial size $x_p \in R^{24 \times (1.W.C)}$, where 24 represents the number of hours (the sequence spans 24 h), C is the number of input channels, W is the number of measurements taken each hour (different metrics like heart rate, blood pressure, etc.).

To process this sequence, we divide it into smaller patches with a size of $(1, W, C)$ for each patch. Each of them contains information for one hour. In total, there are 24 patches, each corresponding to one hour in the sequence. We then flatten these patches, creating a sequence of vectors with size $x_p \in \mathbb{R}^{24 \times (1.W.C)}$ and map them to a higher dimension (Eq. 2) to match the fixed dimension D of the Transformer. This is the process of transforming the original time series data into a sequence of patches that the Transformer model can process.

A learnable embedding vector is added to the embedding sequence, and its state after passing through the Transformer Encoder, denoted as z_L^0 is used to predict the label indicating the patient's survival status y (Eq. 5). For pretraining, the final classification head is an MLP with one hidden layer, while in the fine-tuning phase, it consists of only a single linear layer.

The mentioned Transformer Encoder [8] is a combination of multihead self-attention layer (MSA), multi-layer perceptron layer (MLP) (Eq. 3,4), with each layer being preceded by Layernorm (LN) and followed by residual connections to ensure feature stability.

$$z_0 = \left[x_{class}; x_p^1 E; x_p^2 E; \ldots; x_p^N E \right] + E_{pos}, E \in R^{(W.C) \times D}, E_{pos} \in R^{(N+1) \times D} \quad (2)$$

$$z'_l = MSA(LN(z_{l-1})) + z_{l-1}, \; l = 1 \ldots L \quad (3)$$

$$z_l = MLP\left(LN\left(z'_l\right)\right) + z'_l, \; l = 1 \ldots L \quad (4)$$

$$y = LN\left(z_L^0\right), \quad (5)$$

4 Experimental Setup

4.1 Dataset

MIMIC-III [11] (Medical Information Mart for Intensive Care III) database is a publicly available collection of healthcare data from over 40,000 patients admitted to the Intensive Care Unit (ICU) at Beth Israel Deaconess Medical Center between 2001 and 2012. It includes comprehensive details on patient health conditions, treatment protocols, diagnostic results, and medical histories.

SPHERE dataset [12] was gathered from homes in Bristol, UK, using multiple sensing technologies, including wrist-worn accelerometers, RGB-D cameras (which record both video and depth information), and passive environmental sensors. Annotators manually labeled the data to provide accurate ground-truth labels for human activities such as posture, movement, and location. This dataset is designed to facilitate the creation and testing of algorithms for activity recognition and health monitoring.

4.2 Data Pre-processing

MIMIC-Extract [9] is an open-source code used for extracting, pre-processing and performing data from MIMIC-III. It includes information on demographics during the time of ICU, time-varying vital signs, laboratory test results, and outcomes following hospitalization (survival or death). The approach of MIMIC-Extract involves generating a comprehensive time series dataset suitable for various prediction tasks, while also providing flexible options for dataset cohort and variable selection.

4.3 Implementation Detail

All models mentioned in this research were implemented on the MacBook Pro M1 2020, which features an 8-core CPU for optimized performance, an 8-core GPU for enhanced graphics, and a 16-core Neural Engine for advanced machine learning.

For the Logistic Regression and Random Forest, we used a dictionary distribution (DictDist) to define the hyperparameter space and employed a random search method for sampling.

We trained the ViT model with a hidden embedding size of 128 for all layers of the Encoder, using 6 layers and 8 heads in the multihead self-attention mechanism. The size of the MLP in the feedforward neural network was set to 2048 with 1 input channel. For binary classification, the training process utilized the Cross Entropy Loss function and the Adam optimizer. The model was trained for a total of 20 epochs with a batch size of 64.

The TS-TCC model was configured with the following settings: kernel size of 9, a stride of 1, and 64 output channels, along with a dropout layer with a rate of 0.35. We also trained this model for 15 epochs using the Adam optimizer (Kingma & Ba, 2017) with parameters $\beta_1 = 0.9$ và $\beta_2 = 0.99$. Notably, data augmentation parameters included a Jitter transformation with a rate of 0.001 and a Jitter scaling rate of 0.001, with a maximum number of segments set to 5. Additionally, the hidden dimension size was set to 100 and the number of time steps for the model was 10.

5 Results

(See Table 1).

Table 1. Comparison between results from predictive models

Model	Training time	Accuracy
Logistic Regression	60 min	0,9058
Random Forest		0,9033
TS-TCC	13,33 min	0,877
CNN	3,3 min	0,8975
ViT	**1 min**	**0,8911**

5.1 Logistic Regression and Random Forest

After conducting 15 times random searches to obtain the optimal hyperparameters for the model, the Random Forest achieved an accuracy of **0.9033** and a loss of **0.2489** on the test set, while the Logistic Regression model yielded an accuracy of **0.9058** and a loss of **0.2536**, which is only slightly higher compared to the Random Forest model.

5.2 Time-Series Representation Learning via Temporal and Contextual Contrasting

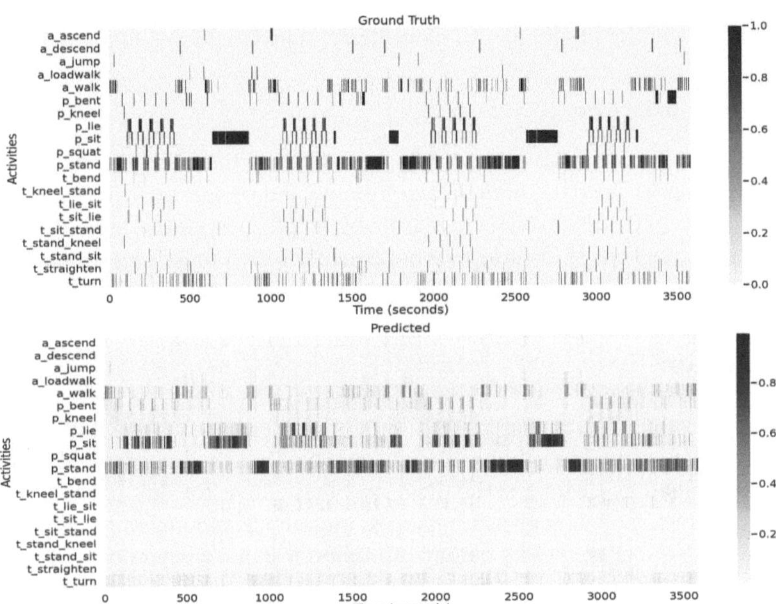

Fig. 4. Predicting actions using SPHERE dataset.

The predictions and its ground truth on the SPHERE dataset shows that this model can predict different types of actions relatively accurately. While the ground truth indicates the probabilities of activities in relatively dark colors, the predictions have paler colors indicating lower probabilities. The activities in the dataset are not uniformly distributed, with some of them appearing many times and others rarely. Therefore, the model cannot exactly predict rare activities compared to more common ones (Fig. 4).

However, the performance on the MIMIC dataset was not satisfactory. Although the loss decreased gradually across epochs, after applying self-supervised learning and fine-tuning for prediction on the test set, there was a sharp drop in loss during the initial epochs. This suggests that the initial parameters of the TS-TCC model did not effectively capture patterns, requiring the model to readjust its weights to better learn the data representations. Consequently, self-supervised learning did not prove meaningful in this context (Fig. 5).

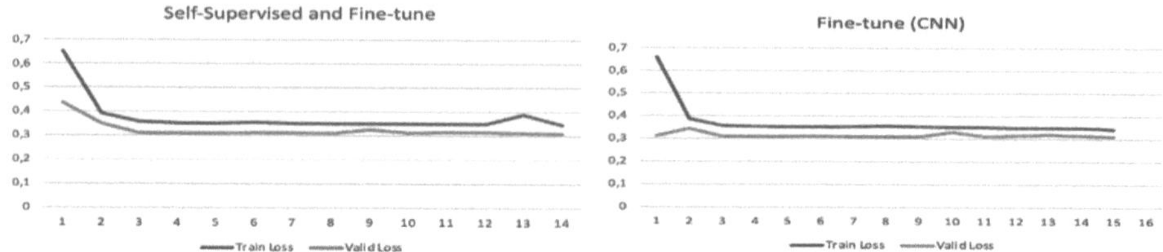

Fig. 5. Results from TS-TCC model and CNN.

The core of fine-tuning involves CNN, and using CNN alone showed a similar trend to the TS-TCC model. Therefore, the TS-TCC model is not well-suited for this type of data. Instead, a simpler model like CNN is appropriate for predicting patient mortality.

5.3 Vision Transformer

The model demonstrates stable performance, indicating that it has effectively learned the characteristics of the data, both during training and on test set, with an accuracy of 0.8911 and a loss of 0.3438, which are reasonably good. Despite this stability, there is no significant improvement in accuracy across epochs, which may suggest that the model has reached the limits of its learning capacity from the current dataset.

6 Conclusion

For the objective of our study, the TS-TCC model was found to be unsuitable. It required significant training time but achieved low accuracy. Thus, self-supervised learning for deeper representations of time series data was ineffective for this patient time series dataset. In contrast, models such as Logistic Regression, Random Forest, CNN, and ViT performed very well in predicting patient mortality, with accuracy rates exceeding 89%. Notably, the ViT model had very short training times per epoch while delivering excellent results. Both Logistic Regression and Random Forest also demonstrated high accuracy; however, a drawback was the time-consuming process of random search to determine hyperparameters for these models. Overall, it can be concluded that this prediction task is relatively straightforward, with performance reaching saturation with minimal data and even with relatively simple models such as Random Forest.

References

1. Che, Z., Purushotham, S., Cho, K., Sontag, D., Liu, Y.: Recurrent Neural Networks for Multi-variate Time Series with Missing Values (2016).arXiv:1606.01865. https://doi.org/10.48550/arXiv.1606.01865
2. Dosovitskiy, A., et al.: An Image is Worth 16x16 Words: transformers for image recognition at scale. In: International Conference on Learning Representations (2020). https://openreview.net/forum?id=YicbFdNTTy
3. Eldele, E., et al.: Time-series representation learning via temporal and contextual contrasting. In: Proceedings of the Thirtieth International Joint Conference on Artificial Intelligence, pp. 2352–2359 (2021). https://doi.org/10.24963/ijcai.2021/324

4. Harutyunyan, H., Khachatrian, H., Kale, D.C., Steeg, G.V., Galstyan, A.: Multitask learning and benchmarking with clinical time series data. Sci. Data **6**(1), 96 (2019). https://doi.org/10.1038/s41597-019-0103-9

5. Kingma, D.P., Ba, J.: Adam: a method for stochastic optimization (2017).arXiv:1412.6980. http://arxiv.org/abs/1412.6980

6. Nestor, B., et al.: Feature Robustness in Non-stationary Health Records: Caveats to Deployable Model Performance in Common Clinical Machine Learning Tasks (2019). arXiv:1908.00690. https://doi.org/10.48550/arXiv.1908.00690

7. Nowroozilarki, Z., Pakbin, A., Royalty, J., Lee, D.K.K., Mortazavi, B.J.: Real-time mortality prediction using MIMIC-IV ICU data via boosted nonparametric hazards. In: 2021 IEEE EMBS International Conference on Biomedical and Health Informatics (BHI), pp. 1–4 (2021). https://doi.org/10.1109/BHI50953.2021.9508537

8. Vaswani, A., et al.: Attention is all you need. Adv. Neural Inf. Process. Syst. **30** (2017). https://proceedings.neurips.cc/paper_files/paper/2017/hash/3f5ee243547dee91fb d053c1c4a845aa-Abstract.html

9. Wang, S., McDermott, M.B.A., Chauhan, G., Hughes, M.C., Naumann, T., Ghassemi, M.: MIMIC-extract: a data extraction, preprocessing, and representation pipeline for MIMIC-III. In: Proceedings of the ACM Conference on Health, Inference, and Learning, pp. 222–235 (2020). https://doi.org/10.1145/3368555.3384469

10. O'Shea, K., Nash, R.: An introduction to convolutional neural networks. Int. J. Res. Appl. Sci. Eng. Technol. (IJRASET) (2015). arXiv.https://arxiv.org/abs/1511.08458

11. Johnson, A.E.W., et al.: MIMIC-III, a freely accessible critical care database. Sci. Data **3**, 160035 (2016). https://doi.org/10.13026/C2XW26

12. Tokin, E., et al.: A multi-sensor dataset with annotated activities of daily living recorded in a residential setting. Sci. Data **10**, 162 (2023). https://doi.org/10.1038/s41597-023-02017-1

Transformer-Based Models for Predicting High-Risk Diabetes in Women Using Tabular Data

Nguyen Tho[1]([envelope])[ID], Quoc-Thông Nguyen[1][ID], Tran Kim Duc[1,2][ID], Ludovic Koehl[2][ID], and Tran Kim Phuc[2][ID]

[1] International Research Institute for Artificial Intelligence and Data Science, Dong A University, Da Nang, Vietnam
`thon@donga.edu.vn`
[2] University of Lille, ENSAIT, ULR 2461 - GEMTEX - Génie et Matériaux Textiles, 59000 Lille, France

Abstract. Diabetes can lead to several long-term severe health complications; more than 199 million women are affected by diabetes, with projections indicating a rise to 313 million by 2040. Feature Tokenizer Transformer (FT-Transformer) is proposed to predict diabetes in 1858 women and evaluate its performance with Tabnet, XGboost, RF, and LR. SHAP and LIME methods are used for interpreting FT-Transformer's performance. FT - Transformer achieved 93.54%, RF (97.31%), XGboost and Tabnet (96.5%), LR (77.42%). FT - Transformer costs more computationally than others; however, it can potentially apply for disease prediction and classification with features containing (text, image, and tabular data) in the medical field. Through SHAP and LIMEthod, an interactive relationship between glucose and pregnancy features in contributing to diabetes prediction was found; glucose is the most predictor of diabetes.

Keywords: Transformer · machine learning · tabular data · diabetes · SHAP · LIME

1 Introduction

Diabetes mellitus (DM) is a metabolic disorder characterized by abnormally high blood glucose levels. It is a chronic condition that directly impacts the pancreas, resulting in the body's inability to manufacture insulin [11]. The main subtypes of diabetes are Type 1 diabetes mellitus (T1DM) and Type 2 diabetes mellitus (T2DM), which often arise from impaired insulin secretion (T1DM) and/or action (T2DM). Currently, more than 199 million women are affected by diabetes, with projections indicating a rise to 313 million by 2040 [15]. The etiology is attributed to several variables, including overweight or obesity, hypertension, and inadequate nutrition.

S. Thomassey et al. (Eds.): RAIDS 2024, LNICST 673, pp. 68–84, 2026.
https://doi.org/10.1007/978-3-032-14055-5_6

Diabetes can lead to several serious long-term health complications, including macrovascular damage affecting large blood vessels in the heart, brain, and legs and microvascular damage impacting small blood vessels in the kidneys, eyes, feet, and nerves. These complications increase the risk of cardiovascular diseases, kidney failure, vision loss, and neuropathy. Effective management of blood sugar levels, regular monitoring, and a healthy lifestyle are crucial in preventing or delaying these complications. Early detection and intervention can significantly improve patient outcomes and quality of life [32,37]. In public health, medical data is vast, including hospital records, patient medical histories, and results of medical examinations. Traditionally, disease diagnosis and doubt rely on a doctor's expertise, sometimes leading to inaccuracies and overlooked data patterns, potentially compromising patient care. Especially in some underprivileged areas, having a doctor with professional knowledge and good medical supplies for diagnosis is also a huge challenge [23].

As a result, there is a critical need for automated systems that can detect early and accurate diabetes diagnoses, while ensuring timely and precise diagnoses and suitable for implementation in communities. Recent advances in Machine Learning (ML) and Deep Learning (DL) have shown promise in improving diabetes risk prediction. Transformer-based models, in particular, have demonstrated exceptional performance in various natural language processing and computer vision tasks. Their ability to handle sequential data and capture complex patterns can potentially be an approach for analyzing tabular data, such as Electronic Health Records (EHRs) and other clinical datasets. The diabetes dataset is tabular data. Despite the potential of transformer-based models, their application in diabetes risk prediction using tabular data remains largely unexplored.

This paper aims to bridge this gap by investigating the effectiveness of transformer-based models in predicting high-risk diabetes using tabular data. We explore using a Feature Tokenizer Transformer (FT-Transformer) to analyze the diabetes Frankfurt dataset from Frankfurt Hospital, Germany, to identify high-risk diabetes patients. Our approach can potentially improve diabetes risk prediction, enable early interventions, and reduce the burden of diabetes on individuals and healthcare systems.

2   Related Work

Many studies used deep learning (DL) algorithms, such as Convolutional Neural Networks (CNN), Deep Belief Network (DBN), Deep Neural Network (DNN), and Multi-Layer Perceptron (MLP) on the Pima Indian diabetes (PID) dataset, that was collected in 1988. This data includes nine attributes and 768 records describing female patients. For this dataset, [28] achieved the highest accuracy of 76.30% from Naive Bayes among Decision Tree, Support Vector Machine (SVM). [16] used ensemble approach (90% of accuary) with explainable AI. [6] used Deep Neural Network (DNN), Support Vector Machine (SVM), and Random Forest (RF) for classification (around 96%). Most of the previous studies are conducted on the PIMA dataset and success on tabular data [14,19]. However, a lack of studies use model-based-transformer architecture in predicting diabetes with tabular data. Recently, with the ubiquitous success of attention-based architecture, some studies have employed attention-like modules for tabular deep learning as well [2,12,29].

In the current trend of developing prediction models based on the transformer approach, there are some transformer models used to classify and predict tabular data. Each model has its strengths and characteristics. [10] used TabPFN as a Prior-Data Fitted Network (PFN) from a trained Transformer that can do supervised classification for small tabular datasets. This model may be suitable for the PIMA dataset that has 768 records *(sample size ¡ 1000)*. [7] use FT-Transformer (Feature Tokenizer Transformer model). [12] proposed Tabtransformer for tabular data.

This study focuses on the FT-Transformer model's performance in diabetes prediction. The model will be compared with other models such as Tabnet, XGboost, Random Forest, and Logistic Regression. To get insight FT - Transformer's performance and important features influencing predictability. Understanding how the model's performance and important attributions for a specific dataset is essential after training and testing models. Explainable Artificial Intelligence techniques have been applied to explain models' performance in diabetes diagnosis, such as interpretable model-agnostic explanation (LIME) and Shapley additive explanation (SHAP). [1,30] used LIME and SHAP analysis for explainable ML-Based Diabetes Predictions.

3   Methodology

3.1   Type 2 Diabetes Dataset

As mentioned, our study uses the Frankfurt dataset[1], which contains a total of 2000 records from 2000 female participants [26]. The values are collected from mobile phone sensor data. There are 8 attributes selected for the dataset, as shown in Table 1. There are 142 patients detected as outliers based on predefined criteria, which are pregnancy ≥ 15 [22], BMI ≤ 51 [27], skin thickness size ≤ 50 [5], these abnormal values are excluded from the dataset, reducing the number of patients to 1,858. The variables where zero values were considered invalid or missing (such as Glucose, Blood Pressure, Skin Thickness, Insulin, and BMI) are summarized in Table 2. The imputation was performed using the mean values of each respective variable. This step ensured that all necessary attributes had valid data for analysis and modeling. Finally, the processed dataset of 1,858 patients was split into training and test sets. The training set contains 1,486 patients (80% of the data), and the test set has 372 patients (20% of the data). The following steps to prepare the data for model training are demonstrated in Fig. 1.

A sequence of data analysis used for diabetes prediction is depicted in Fig. 2. The process begins with the raw dataset undergoing preprocessing before being

---

[1] https://ieee-dataport.org/documents/type-2-diabetes-dataset.

Table 1. The attributes of Frankfurt dataset

Attribute	Description	Type	Average/Mean ± SD
Pregnancies	Number of times pregnant.	Numeric	3.7 ± 3.3
Glucose	Plasma glucose concentration 2h in an oral glucose tolerance test.	Numeric	121.18 ± 32.07
Blood pressure	Diastolic blood pressure (mm Hg).	Numeric	69.14 ± 19.18
Skin Thickness	Triceps skinfold thickness (mm).	Numeric	20.93 ±16.1
Insulin	2-hour serum insulin (μ IU/mL).	Numeric	80.25 ± 111.18
BMI	Body mass index (kg/m^2).	Numeric	32.19 ±8.15
DPF	Diabetes pedigree function.	Numeric	0.47 ± 0.32
Age	Age (years).	Numeric	33.09 ± 11.79
Outcome	Diabetes diagnose results (tested_positive: 1, tested_negative: 0)	Nominal	–

Table 2. The number of missing values in the Frankfurt dataset.

Attributes	No. of missing values
Pregnancies	0
Glucose	13
Blood Pressure	90
Skin Thickness	573
Insulin	956
BMI	28
DPF	0
Age	0

split into training and testing sets, with 80% of the data allocated for training (n = 1486) and 20% for testing (n = 372). Then, FT-Transformer, Tabnet, Xgboost, Random Forest, and Logistic Regression models are trained to predict diabetes. These models are compared and evaluated based on their performance on the test data. If the model performance is unsatisfactory, adjustments to parameters such as epoch, batch size, and learning rate are made, and the process is repeated. Once an optimal model is identified, local explanation methods such as SHAP and LIME are applied to the FT-Transformer model to provide interpretability.

3.2   Feature Tokenizer Transformer Architecture

FT-Transformer (Feature Tokenizer Transformer) is a simple adaptation of the Transformer architecture for tabular domain [7,34]. In this study, we focus on the FT-Transformer model, the proposed FT-Transformer model architecture in Fig. 3. FT-Transformer transforms attributions (categorial and numeric) to embeddings and applies a stack of Transformer layers. Thus, every Transformer layer operates on the feature level of one object. Variables having zero mutual

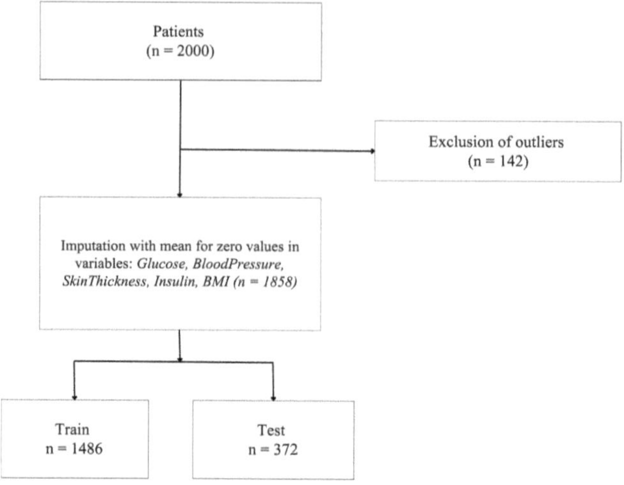

Fig. 1. Flow chart on the pre-processing of data

information score are excluded. There are not zero mutual information score variables. The proposed FT-Transformer is summarized in the Table 3.

Table 3. FT-Transformer Model Summary

Component	Description
Embedding Dimension	128
Encoder Layers	2 layers
Attention Heads	4 attention heads
Feedforward Dimension	256
Output Layer	Linear layer with 1 output unit for binary classification.
Activation	Sigmoid activation for binary classification.

3.3   Tabnet

TabNet is a deep learning architecture built by the Google Cloud team in the year 2019 for tabular data. It utilizes sequential attention mechanisms with an encoder to choose relevant features at each decision step selectively [2]. Recently, in the medical field, Tabnet has been used for spinal cord tumors [17], Parkinson's disease [18], Alzheimer's disease [13].

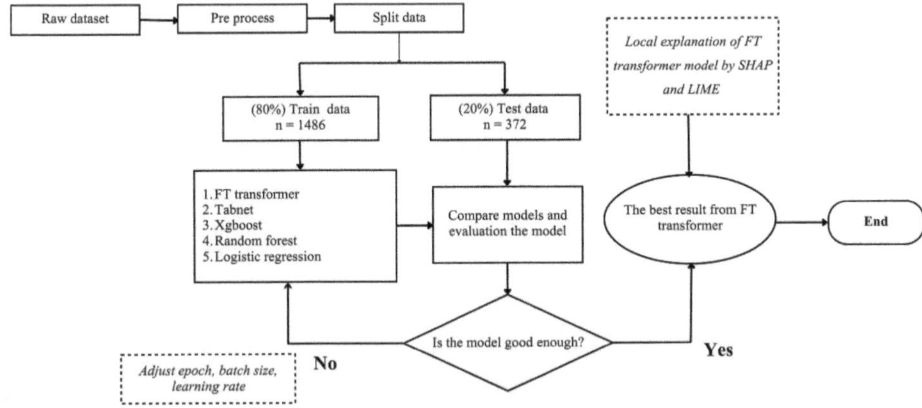

Fig. 2. The flow diagram outlining the sequence data analysis process for diabetes prediction

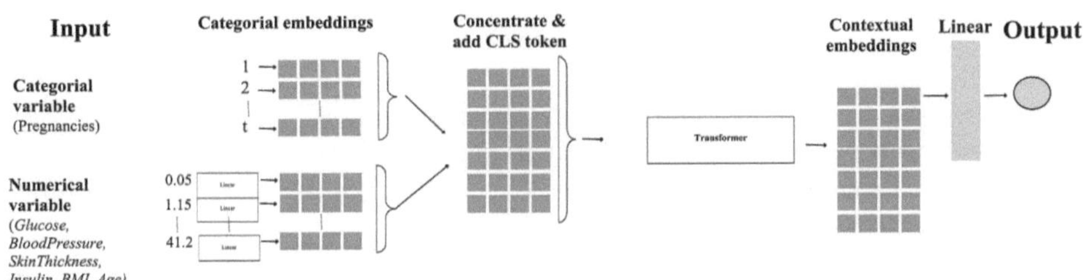

Fig. 3. FT-Transformer Model Architecture

3.4   eXtreme Gradient Boosting (XGboost)

XGBoost is an advanced machine learning algorithm that falls under the ensemble learning category, utilizing the gradient boosting framework. It leverages decision trees as its fundamental building blocks and incorporates regularization methods to improve model generalization and prevent overfitting. XGBoost algorithm shows computational efficiency, provides insightful analysis of feature importance, and effectively manages missing data. It is widely used for regression, classification, and ranking in diabetes classification [9,36].

3.5   Random Forest

Random forest (RF) is a supervised machine-learning classification and regression analysis technique. It produces great results without hyperparameter adjustment [3]. Due to its simplicity and versatility, RF is often utilized in the medical domain to classify diseases, such as diabetes [21,35].

3.6   Logistic Regression

Logistic regression (LR) is a statistical technique used to examine the relationship between independent variables and a categorical outcome. It establishes a boundary, or hyper-plane, that separates two distinct groups within the data. It defines a functional form and a set of parameters representing the probability of a particular outcome based on the input variables. LR is often used in medicine to predict the risk of developing a disease or classify data into different categories, such as diabetes [24].

3.7   Feature Impact Explanations

Explainable artificial intelligence is beneficial in converting opaque machine learning models into transparent ones and outlining how each one makes decisions in the healthcare industry. To comprehend the variables that affect decision-making regarding diabetes prediction that can be accounted for by model-agnostic techniques [1].

SHapley Additive exPlanations values, a robust framework within Explainable AI (XAI), provide a precise understanding of how each feature contributes to a model's predictions. By assigning an importance score to each feature, SHAP quantifies its influence on the output, with higher scores reflecting a more significant impact on the predicted outcome and lower scores indicating a minimal effect [8].

Interpretation Local Interpretable Model-agnostic Explanations provide interpretability at the level of individual predictions, focusing on the specific factors and feature interactions that led to a particular conclusion. For example, in predicting that a patient does not have diabetes, LIME delves deeply into these factors, offering a granular view of the underlying decision-making process [30].

3.8   Performance Evaluation

A confusion matrix is utilized to assess the performance of models in binary classification tasks. The confusion matrix encapsulates the anticipated and actual classifications, enabling researchers to compute multiple assessment measures that elucidate the model's performance. The findings require analysis as they assist in establishing the model's threshold and fine-tuning hyperparameters for optimal performance. A confusion matrix illustrates the quantities of true positives (TP), false positives (FP), true negatives (TN), and false negatives (FN) given a series of forecasts with their actual ground truth labels. Alongside the previously stated assessment metrics, various scholars have undertaken further research to assess model performance utilizing statistical curves, including the Receiver Operating Characteristic (ROC) curve and the Area Under the Curve (AUC).

Accuracy is the proportion of correct predictions to the total number of predictions made, calculated as

$$\text{Accuracy} = \frac{TP + TN}{TP + TN + FP + FN}.$$

Precision. The proportion of true positives to all positive predictions made by the model.

$$\text{Precision.} = \frac{TP}{TP + FP}.$$

Recall is the proportion of true positives to all positive samples in the data,

$$\text{Recall.} = \frac{TP}{TP + FN}.$$

F1 Score computes the harmonic mean of precision and recall, which provides a balance between the two,

$$\text{F1 Score} = 2 \times \frac{\text{Precision} \times \text{Recall}}{\text{Precision} + \text{Recall}}.$$

Specificity presents the proportion of true negatives to all negative samples in the data,

$$\text{Specificity} = \frac{TN}{TN + FP}$$

Balanced Accuracy is computed as follow

$$\text{Balanced Accuracy} = \frac{1}{2} \left(\frac{TP}{TP + FN} + \frac{TN}{TN + FP} \right).$$

In this study, we initially find the better FT transformer compared to different learning rates and epochs of each model. After achieving the highest accuracy model, we choose those parameters. Finally, we focus on the relative performance of different architectures with other models (Tabnet, Xgboost, Random forest, Logistic Regression).

4   Results

Table 4 has shown the FT-Transformer model reaches the highest accuracy to 97.31% when training with learning rate = 0.001, epochs = 1000, and batch size = 128. The model's accuracy decreased slightly from 0.001 to 0.001 learning rate. However, the accuracy declined substantially by 12% for learning rate (0.00001). The model with a learning rate (0.001) was chosen to train the model in the next step, in which epochs were changed at 200, 400, and 800 epochs.

At the Table 5, the trained model with 800 epochs has shown the highest accuracy (97.58%) in the test dataset. This model's performance is slightly higher

Table 4. Impact of learning rate on accuracy measurement of FT-Transformer at 1000 epochs, batch size = 128

Learning rate	Accuracy
0.001	**97.31%**
0.0001	96.51%
0.00001	84.68%

Table 5. At learning rate 0.001 with 02 hidden layer changes impact on the accuracy.

Epochs	Accuracy	Training accuracy	Testing accuracy
200	0.9914	100 %	96.24%
400	0.9914	99.93%	95.97%
800	0.9952	100%	**97.58%**

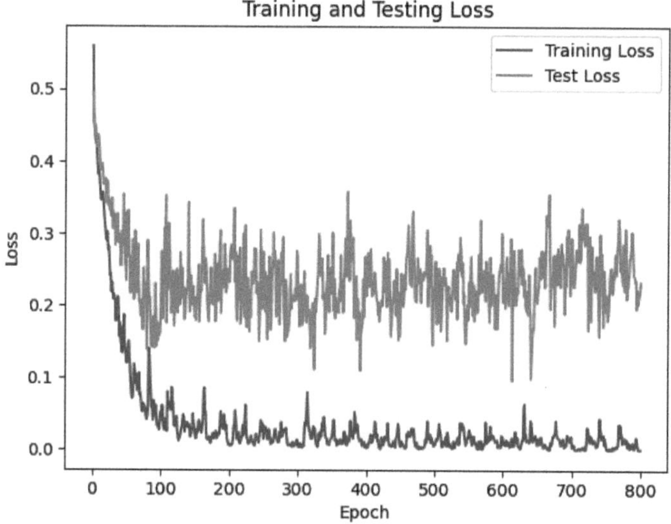

Fig. 4. History of training, test data from FT-Transformer model with 800 epochs, 0.001 learning rate

than the model with 1000 epochs, and the model trained 800 epochs mitigate the training time and computational costs and may avoid overfitting for test dataset prediction [25]. Therefore, the FT-Transformer with two hidden layers, learning rate (0.001), batch size (128), and trained 800 epochs was opted to compare with Tabnet, Xgboost, RF, and LR models(Figs. 4,5 and 6).

From the confusion matrix in Fig. 7, the FT-Transformer's performance reaches 97% in accuracy prediction, sensitivity (100%), and specificity (96.6%).

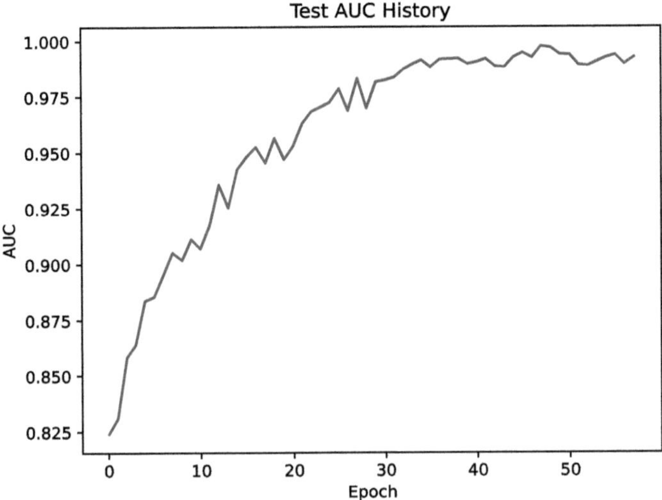

Fig. 5. Test AUC history of FT-Transformer model

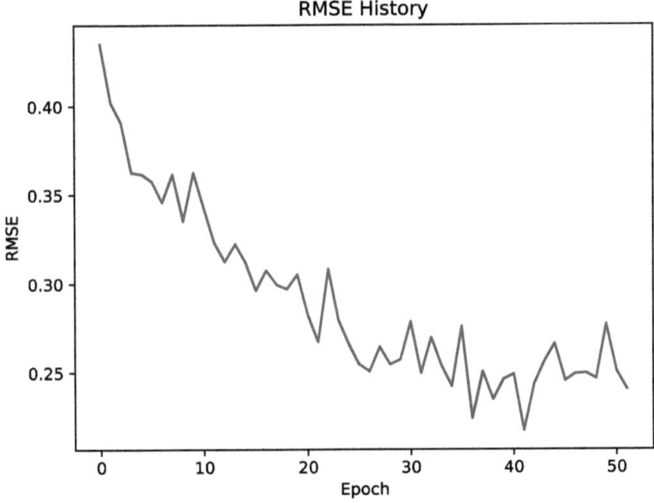

Fig. 6. Root Mean Square Error of FT-Transformer model

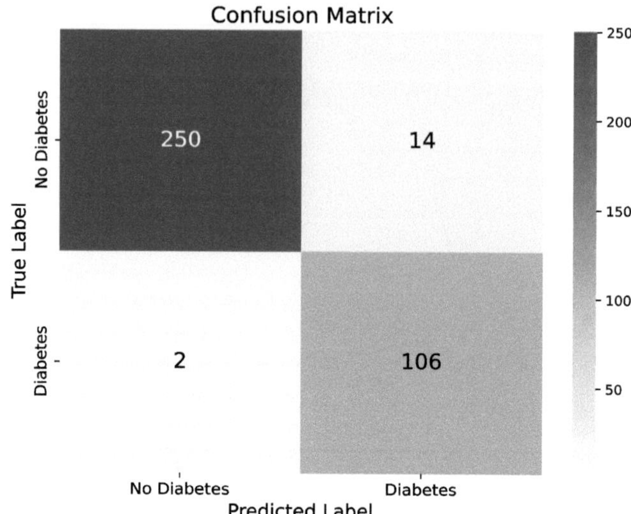

Fig. 7. Confusion matrix for the FT-Transformer model

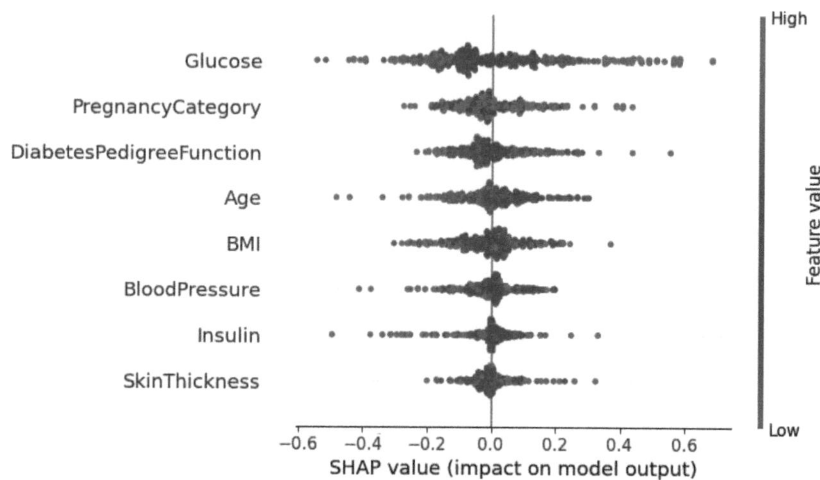

Fig. 8. SHAP summary plot for the FT-Transformer model

4.1 SHapley Additive exPlanations

Figure 8 has shown that "Glucose" and "Pregnancy times" are the most influential features for diabetes prediction. At the same time, "Blood Pressure" and "Skin Thickness" do not significantly impact the model's prediction.

Figure 9 displays the "Glucose" and its corresponding SHAP values, and its interaction with "Pregnancy". Particularly, there is a significant statistical correlation among glucose levels, pregnancy times, and diabetes. Higher glucose

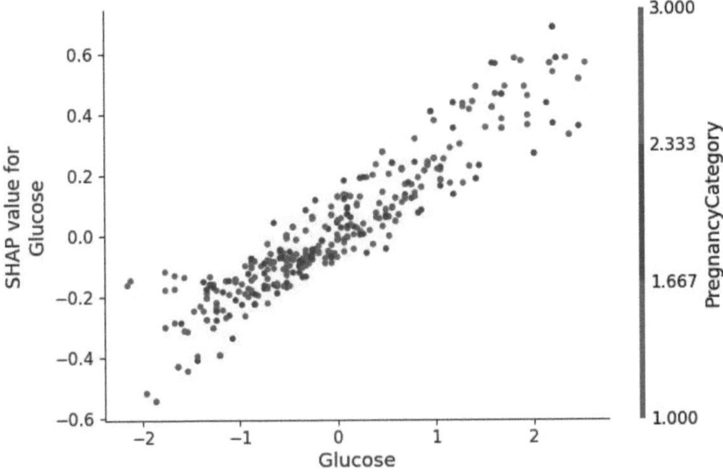

Fig. 9. SHAP dependence plot

levels are positively correlated with the prediction of diabetes. Although pregnancy times also impact moderately on the prediction of diabetes, glucose is still the dominant predictor.

4.2 Interpretation Local Interpretable Model-Agnostic Explanations (LIME)

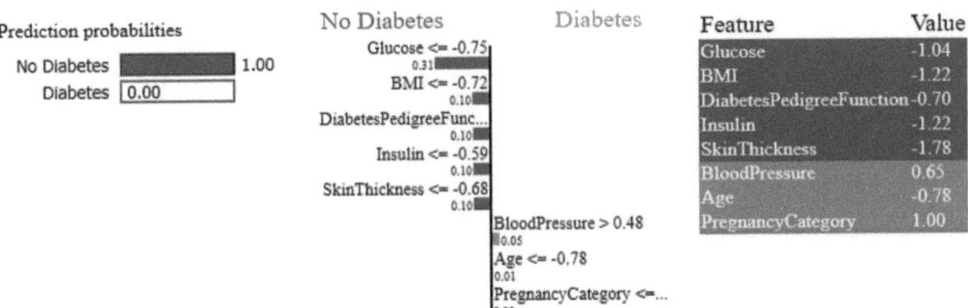

Fig. 10. Local Interpretable Model-agnostic Explanations for FT-Transformer on Frankfurt dataset

Figure 10 has shown that While certain features (high Diabetes Pedigree Function and Insulin levels) suggest a tendency towards diabetes, other significant features such as BMI, Glucose, and Age strongly favor "No Diabetes". Patients who have lower BMI $\leq$-0.70 (0.21) contribute to predicting "No Diabetes", similar -0.68 < Glucose $\leq$ -0.12 (0.17) predict "No diabetes". Pregnancies, blood pressure, and thickness have less influence on prediction. Diabetes Pedigree Function has a significant impact on prediction, however, the insulin variable has a minor impact.

Table 6. Performance measure of all classification methods for Test data

Classification	Precision	Recall	F-measure	Accuracy
Logistic Regression	0.622	0.565	0.592	77.42%
Random Forest	0.945	0.963	0.954	**97.31%**
XGboost	0.906	0.981	0.942	96.50%
Tabnet	0.906	0.981	0.942	96.50%
FT-Transformer	0.882	0.898	0.890	**93.54%**

At the Table 6, FT-Transformer (93.54%) and Random Forest (97.31%) are the best overall models for Frankfurt dataset with balancing precision, recall, and overall accuracy.

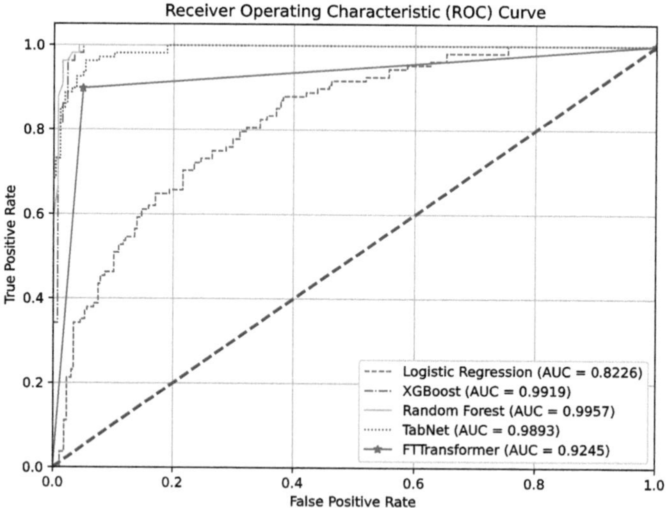

Fig. 11. ROC AUC of FT-Transformer, Tabnet, XgGboost, Random Forest and Logistic Regression

From Fig. 11, Machine learning models achieved the higher ROC AUC values than deep learning models (except for Logistic Regression), while Random Forest's AUC values is highest point (0.9957), XGboost (0.9919), Tabnet (0,989) and FT - Transformer (0,924).

5   Challenges

Based on data processing and findings, there are two main points for challenges. Firstly, medical data is predominantly in tabular format and big and complex electronic health records. At the same time, almost all current ML and AI models focus on image and text data [33]. Therefore, developing tabular foundation models and comparing current ML and AI models for tabular data is essential to finding effective models for processing and analyzing massive datasets. Secondly, collecting data traditionally may cause bias. The Frankfurt diabetes dataset is an example. There are approximately 10% of large outlier points and a range of missing data at some variables from 5% to 30%, which may cause bias and high dimensionality in training models. Using digital collecting data with conditional characteristics for some features can reduce data bias and enhance the quality of data for the training process [4,31].

6   Discussion

FT-Transformer model's performance is excellent, however it is not higher than other machine learning models (RF, XGboost) and deep learning (Tabnet), except for LR in the Frankfurt dataset (1858 records); Transformer model may not be the best method for tabular data. This finding can be understood that tabular data (medical area) usually has heterogeneous features (continuous, categorical, ordinal values,..). Moreover, the Transformer model was designed mainly for text and images, that tokens (characters and pixels) are ordered to have semantic meaning and correlation [34], while the contrast is true for tabular data. Regarding deep tabular data learning architectures with attention mechanisms (Tabnet and Transformer). In this study, for test dataset, the Tabnet model has shown outperformance in prediction by saving computational costs (0.000039 s), which is faster than FT - Transformer and machine learning models.

In comparison with previous studies in the same dataset (Frankfurt) and PIMA dataset for diabetes prediction. Random Forest is still one of the better performance, at Frankfurt dataset, [6] (96% of accuracy), at PIMA dataset [20] 97%, [4] 86%. Although FT-Transformer costs computation, when applying the model in a medical clinical support system, the model can work effectively on big and complex data in qualitative (sequential) and quantitative (tabular) analysis. In general, models based-transformer architecture promises potential advantages in the medical area,

In the development trend of electric diagnosis systems and smart healthcare, the study can contribute compelling evidence for selecting appropriate models to embed into the system. This application can support doctors in monitoring remotely and progress of patients' diseases (diabetes, heart diseases).

7   Conclusion

The study's purpose is to assess FT-Transformer models with Tabnet (attention mechanism) and machine learning (XGboost, RF, LR). RF, XGboost and Tabnet show outperformace with FT - Transformer.. Using SHAP and LIME support to explain the model's performance on predicting risk factors (important features) associated to diabetes.

In our future works, the Frankfurt dataset is not large, has fewer features for prediction, and is only specific for female German people, therefore, in the next phase, to have a more accurate assessment and applicability in Vietnam or Asia, we consider to collect Vietnamese data from hospitals with more features and large data. Furthermore, developing and experimenting with a model-based transformer for detecting and predicting diabetes and prognosis over time is crucial.

References

1. Ahmed, S., Kaiser, M. S., Hossain, M. S., Andersson, K.: A comparative analysis of lime and shap interpreters with explainable ml-based diabetes predictions. IEEE Access (2024)
2. Arik, S, Ö., Pfister, T.: Tabnet: attentive interpretable tabular learning. In: Proceedings of the AAAI Conference On Artificial Intelligence, pp. 6679–6687 (2021)
3. Arowolo, M. O., Aigbogun, H. E., Michael, P. E., Adebiyi, M. O., Tyagi, A. K.: A predictive model for classifying colorectal cancer using principal component analysis. In: Data Science for Genomics, pp. 205–216. Elsevier (2023)
4. Chang, V., Bailey, J., Xu, Q.A., Sun, Z.: Pima indians diabetes mellitus classification based on machine learning (ml) algorithms. Neural Comput. Appl. **35**(22), 16157–16173 (2023)
5. Derraik, J.G., et al.: Effects of age, gender, bmi, and anatomical site on skin thickness in children and adults with diabetes. PLoS ONE **9**(1), e86637 (2014)
6. Bouhissi, H. E., et al.: Towards diabetes mellitus prediction based on machine-learning. In: 2023 International Conference on Smart Computing and Application (ICSCA), pp. 1–6. IEEE (2023)
7. Gorishniy, Y., Rubachev, I., Khrulkov, V., Babenko, A.: Revisiting deep learning models for tabular data. Adv. Neural. Inf. Process. Syst. **34**, 18932–18943 (2021)
8. Guleria, P., Srinivasu, P.N., Hassaballah, M.: Diabetes prediction using shapley additive explanations and dsaas over machine learning classifiers: a novel healthcare paradigm. Multimedia Tools and Applications **83**(14), 40677–40712 (2024)
9. Gündoğdu, S.: Efficient prediction of early-stage diabetes using xgboost classifier with random forest feature selection technique. Multi. Tools Appl. **82**(22), 34163–34181 (2023)
10. Hollmann, N., Müller, S., Eggensperger, K., Hutter, F.: Tabpfn: a transformer that solves small tabular classification problems in a second. arXiv preprint arXiv:2207.01848 (2022)
11. Holt, R. I., Flyvbjerg, A.: Textbook of diabetes. John Wiley and Sons (2024)
12. Huang, X., Khetan, A., Cvitkovic, M., Karnin, Z.: Tabtransformer: tabular data modeling using contextual embeddings. arXiv preprint arXiv:2012.06678 (2020)
13. Jin, Y., et al.: Classification of alzheimer's disease using robust tabnet neural networks on genetic data. Math. Biosciences Eng. MBE **20**(5), 8358–8374 (2023)
14. Kadra, A., Lindauer, M., Hutter, F., Grabocka, J.: Well-tuned simple nets excel on tabular datasets. Adv. Neural. Inf. Process. Syst. **34**, 23928–23941 (2021)
15. Kapur, A., Seshiah, V.: Women and diabetes: our right to a healthy future (2017)

16. Kibria, H.B., Nahiduzzaman, M., Goni, M.O.F., Ahsan, M., Haider, J.: An ensemble approach for the prediction of diabetes mellitus using a soft voting classifier with an explainable ai. Sensors **22**(19), 7268 (2022)
17. Kita, K., et al.: Bimodal artificial intelligence using tabnet for differentiating spinal cord tumors—integration of patient background information and images. Iscience, 26(10) (2023)
18. Kumar, T., Ujjwal, R.: Tabnet unveils predictive insights: a deep learning approach for parkinson's disease prognosis. International Journal of System Assurance Engineering and Management, pp. 1–10 (2024)
19. Meijerink, L., Cinà, G., Tonutti, M.: Uncertainty estimation for classification and risk prediction on medical tabular data. arXiv preprint arXiv:2004.05824 (2020)
20. Naz, H., Ahuja, S.: Deep learning approach for diabetes prediction using pima indian dataset. J. Diabetes Metabolic Disorders **19**, 391–403 (2020)
21. Ooka, T., Johno, H., Nakamoto, K., Yoda, Y., Yokomichi, H., Yamagata, Z.: Random forest approach for determining risk prediction and predictive factors of type 2 diabetes: large-scale health check-up data in japan. BMJ Nutrition, Prevention Health **4**(1), 140 (2021)
22. Prütz, F., Hintzpeter, B., Krause, L.: Abortions in germany-current data from the statistics on terminations of pregnancy. J. Health Monitoring **7**(2), 39 (2022)
23. Rahman, K. M. T., et al.: Challenges faced by medical officers in providing healthcare services at upazila health complexes and district hospitals in bangladesh–a qualitative study. The Lancet Regional Health-Southeast Asia, 24 (2024)
24. Rajendra, P., Latifi, S.: Prediction of diabetes using logistic regression and ensemble techniques. Comput. Meth. Prog. Biomed. Update **1**, 100032 (2021)
25. Rawat, R., Patel, J.K., Manry, M. T.: Minimizing validation error with respect to network size and number of training epochs. In The 2013 international joint conference on neural networks (IJCNN), pp. 1–7. IEEE (2013)
26. Rupabanta Singh, K.: Type 2 diabetes dataset (2024). https://dx.doi.org/10.21227/xm4p-nx87
27. Schienkiewitz, A., Kuhnert, R., Blume, M., Mensink, G.B.: Overweight and obesity among adults in germany-results from geda 2019/2020-ehis. J. Health Monitoring **7**(3), 21 (2022)
28. Sisodia, D., Sisodia, D.S.: Prediction of diabetes using classification algorithms. Proc. comput. sci. **132**, 1578–1585 (2018)
29. Somepalli, G., Goldblum, M., Schwarzschild, A., Bruss, C. B., Goldstein, T.: Saint: improved neural networks for tabular data via row attention and contrastive pre-training. arXiv preprint arXiv:2106.01342 (2021)
30. Tasin, I., Nabil, T.U., Islam, S., Khan, R.: Diabetes prediction using machine learning and explainable AI techniques. Healthcare Tech. Lett. **10**(1–2), 1–10 (2023)
31. Thi, L. M., Tho, N., Phuong, B. T., Mai, V. T. T., Huong, N. T.: How kobotoolbox versus unipark platform are selected in data survey to detect child maltreatment in vietnam: A discussion paper. Journal of Health and Development Studies, 7(4) (2023)
32. Trikkalinou, A., Papazafiropoulou, A.K., Melidonis, A.: Type 2 diabetes and quality of life. World J. Diabetes **8**(4), 120 (2017)
33. van Breugel, B., van der Schaar, M.: Why tabular foundation models should be a research priority. arXiv preprint arXiv:2405.01147 (2024)
34. Vaswani, A.: Attention is all you need. Advances in Neural Information Processing Systems (2017)

35. VijiyaKumar, K., Lavanya, B., Nirmala, I., Caroline, S. S.: Random forest algorithm for the prediction of diabetes. In: 2019 IEEE international conference on system, computation, automation and networking (ICSCAN), pp. 1–5. IEEE (2019)
36. Xu, Z., Wang, Z.: A risk prediction model for type 2 diabetes based on weighted feature selection of random forest and xgboost ensemble classifier. In: 2019 eleventh international conference on advanced computational intelligence (ICACI), pp. 278–283. IEEE (2019)
37. Zakir, M., et al.: Cardiovascular complications of diabetes: from microvascular to macrovascular pathways. Cureus, 15(9), 2023

Explainable Transformer-Based Approach for ECG Anomaly Detection

Thi Thuy Van Nguyen[1,2(✉)] [iD], Cédric Heuchenne[1] [iD], Kim Duc Tran[3] [iD], Guillaume Tartare[2] [iD], and Kim Phuc Tran[2] [iD]

[1] HEC Liège - Management School, University of Liège, Liège, Belgium
{ttv.nguyen,c.heuchenne}@uliege.be
[2] GEMTEX, ENSAIT, University of Lille, Lille, France
{guillaume.tartare,kim-phuc.tran}@ensait.fr
[3] IAD, Dong A University, Da Nang, Vietnam
ductk@donga.edu.vn

Abstract. Detecting anomalies is critical in various fields, especially in healthcare. The ability to identify abnormal patterns from normal ones can help clinicians make early interventions and improve patient outcomes. In Electrocardiogram (ECG) analysis, timely detection of unusual signals is crucial for diagnosing and treating serious, life-threatening health problems. However, although many AI-based anomaly detection models offer high performance, they often function as "black boxes", making us difficult to interpret the results. In this paper, we propose integrating explainable artificial intelligence (XAI) into a transformer-based network combined with a Support Vector Data Description control chart and multivariate exponential weighted moving average technique (MEWMA-SVDD chart) to obtain a robust ECG anomaly detection model. By incorporating XAI, we aim to enhance the transparency and reliability of our model, providing clear and interpretable results. We will demonstrate our approach's effectiveness by using a well-known ECG dataset and provide important insights into the detection mechanism. This approach illustrates the importance of combining advanced deep learning techniques with XAI to improve the reliability and efficiency of anomaly detection systems in monitoring healthcare.

Keywords: Explainable artificial intelligence · Anomaly detection · Transformer · Variational autoencoder · Control chart · Support vector data description · Multivariate exponentially weighted moving average

1 Introduction

Detecting anomalies is vital across various fields since it reveals unexpected deviations from normal patterns. Addressing these irregularities in time will help us improve safety, efficiency, and overall well-being. It is especially important in critical fields such as healthcare, where early detection of anomalies can help to timely intervention and increase the chances of survival for patients. One of the

S. Thomassey et al. (Eds.): RAIDS 2024, LNICST 673, pp. 85–108, 2026.
https://doi.org/10.1007/978-3-032-14055-5_7

most important applications of anomaly detection in healthcare is Electrocardiogram (ECG) monitoring, which provides vital insights for assessing cardiac health. Early detection of anomalies in ECG signals can help to prevent serious conditions like heart attacks and arrhythmias, stroke, and sudden cardiac arrest. Therefore, accurate and timely analysis of ECG data is essential for effective cardiac care (see Amini et al. [4]; Muzammil et al. [23]).

Over the years, various methodologies have been developed to detect anomalies, from traditional statistical approaches, theories from cognitive psychology to advanced machine learning (ML) techniques. Control charts, a fundamental tool in Statistical Process Control (SPC), offer us the advantage of interpretability and have been widely used for anomaly detection in healthcare and other fields (Suman and Prajapati [36]; Lang [15]; Zhou and Kan [43]). However, these methods often struggle with the complexity and high dimensionality of modern datasets. Besides SPC, cognitive psychology also plays an important role in anomaly detection by providing insides into human observational limitations. Theories such as inattentional blindness and the tunnelling effect explain why crucial details are often overlooked when the focus is too narrow or misdirected. To overcome this, training programs and optimized design principles have been implemented to improve detection rates in environments relying on human monitoring. These strategies enhance the performance by ensuring the crucial signs are more likely to be noticed (see Kreitz et al. [14]; Gao and Jia [10]; Saint-Lot et al. [29]). However, despite these improvements, such non-automated methods still face challenges in consistency and scalability, especially when handling large and complex datasets. Conversely, deep learning (DL) models have demonstrated their impressive accuracy in identifying anomalies through the use of large and complex datasets with sophisticated algorithms (Bolhasani et al. [6]; Nguyen et al. [24]). However, despite their effectiveness, these advanced models are often difficult to understand and provide little clarity regarding their decision-making processes. The lack of transparency in these AI-based models leads to significant challenges, especially in vital areas such as health care. In these fields, understanding how a model makes predictions is also as important as the predictions themselves. For example, in ECG monitoring, a "black box" model may correctly detect an anomaly, but it may not provide the necessary explanation for why the anomaly was flagged. This lack of interpretability can make it difficult for healthcare professionals to trust the model's results and make the right decisions. As a result, it may slow down medical decision-making processes and finally affect patient outcomes. Therefore, improving the transparency in AI-based models is crucial to building trust, increasing efficiency, and achieving better healthcare outcomes.

To address this issue, Explainable AI (XAI) offers a solution to improve the clarity of AI models and the transparency of their predictions. The techniques used in XAI aim to provide clear and understandable explanations of how models reach their conclusions, enhance user trust, and enable better decision-making. To date, many widely used methods have been developed to improve the transparency of AI models, with a strong focus on improving their explainability.

One commonly used method is Local Interpretable Model-Agnostic Explanations (LIME), which locally simplifies complex models with simpler, easier-to-understand models to explain individual predictions (Ribeiro et al. [28]). Another popular technique is SHapley Additive exPlanations (SHAP), which uses game theory to assign a significance value to each feature for a particular prediction, providing a consistent and globally interpretable explanation model (Lundberg and Lee [20]). Gradient-weighted Class Activation Mapping (Grad-CAM) is also a frequently employed technique for creating visual explanations, particularly with convolutional neural networks (CNNs), as it identifies the important areas in the input image that contribute to the prediction of the model (Selvaraju et al. [30]). In addition to these post-hoc methods, XAI-by-design has also received attention from researchers recently. XAI-by-design focuses on creating models that are naturally easy to interpret by designing them with transparent structures and understandable decision-making processes. A prime example in this category is the attention mechanism in neural networks, particularly transformers. This mechanism highlights the specific parts of the input data that influence the predictions of the model, providing straightforward explanations of how the model works (Vaswani et al. [40]). In healthcare, the use of XAI methods has greatly enhanced the interpretability of DL models, leading to remarkable improvements. For instance, in a recent study, Prendin et al. [26] used SHAP to assess the effect of different features on glucose level predictions in Type 1 Diabetes management, showing that when embedded in a decision support system, this helped to make better treatment decisions and blood sugar control. Alabi et al. [2] used SHAP and LIME to analyze how different factors influence survival predictions for nasopharyngeal cancer, providing personalized insights into risk factors for each patient. In [25], the authors incorporated Grad-CAM into their deep transfer learning algorithm to interpret the detected COVID-19 cases using CT-scan images and X-rays. The work by Albahri et al. [3] provided a detailed overview of XAI techniques used in healthcare, including methods like SHAP, LIME, Grad-CAM, and various data fusion techniques. The review highlights the crucial importance of transparency and reliability in AI applications. Therefore, integrating XAI into healthcare models is essential for improving clinical decision-making and building trust among healthcare professionals and patients.

In this study, we will construct an XAI framework for ECG anomaly detection. Specifically, we will integrate an explainability module into our previously developed transformer-based network combined with a MEWMA-SVDD chart (Nguyen et al. [24]). This hybrid model was designed to detect any type of anomalies that differ from the normal ECG patterns, making it a flexible tool for detecting a wide range of heart irregularities. The model leverages the strengths of DL for feature extraction and the precise monitoring capabilities of control charts to enhance sensitivity to small changes in the time series data, reduce false alarm rates, and make it highly effective for detecting anomalies in ECG signals. We choose SHAP for its ability to provide detailed, transparent explanations at both global and individual levels, which are crucial for addressing complex data such as ECG. Specifically, we will apply the SHAP method to analyze reconstruction error vectors obtained from the variational autoencoder (VAE) model to uncover which specific window time steps in the ECG segments significantly impact the reconstruction error. At the global level, we will identify which windows contribute the most to the overall model, while at the individual level, we will determine which windows have the most impact on specific ECG instances. This method not only enhances our understanding of the model's performance by revealing the most influential windows in ECG signals but also improves the decision-making process in clinical settings. The remaining work is structured as follows: Sect. 2 provides an overview of the related work and foundational concepts of transformers, VAE, MEWMA-SVDD algorithms, and SHAP principles. Section 3 reviews the detailed architecture of the explainable transformer-based network combined with the MEWMA-SVDD control chart, and then presents the integration of SHAP for enhanced transparency. Section 4 outlines the experiment on a real ECG dataset for evaluating the framework's performance, presents obtained results, and explains the model's interpretability through the integration of SHAP. Section 5 provides the conclusion of the work.

2 Background

In this section, we review some recent developments in ECG monitoring using DL techniques and the main background of our proposed framework, including the transformer architecture, VAE model, MEWMA-SVDD control chart, and SHAP principles.

2.1 Related Work

ECG monitoring is vital for the diagnosis and management of cardiac conditions, and the application of DL techniques has significantly advanced this field. Recently, many innovative approaches have been developed to enhance ECG monitoring with DL. In a study by Liu et al. [17], the authors utilized a vector quantized VAE to address the challenge of insufficient positive samples by employing data augmentation, improving ECG data classification performance. Shin et al. [32] proposed an enhanced AnoGAN model for arrhythmia detection, addressing data imbalance issues and demonstrating improved performance through fine-tuned decision boundaries and learning counts. Jang et al. [12] explored an unsupervised learning approach using a convolutional VAE to extract features from ECG data, showing its utility in anomaly detection and transfer learning for arrhythmia classification. Gaudilliere et al. [11] applied transformer models, typically used in natural language processing, to ECG data, achieving notable success in multi-label classification of heartbeat abnormalities. This demonstrates the versatility and strength of transformer architectures in handling sequential data like ECG signals. The advantages of using transformers include their capacity to effectively capture long-range dependencies and their scalability, which is particularly beneficial for analyzing complex ECG data. In 2023, Raza et al. [27] proposed AnoFed, a hybrid model combining transformer-based VAE and SVDD for monitoring ECG. This model enhances privacy protection and demonstrates high performance in detecting anomalies in ECG signals. After that, Nguyen et al. [24] expanded this model by combining this transformer-based network with a MEWMA-SVDD chart for ECG monitoring. This model utilizes the capabilities of feature extraction of transformer-based VAE network, with the inclusion of the MEWMA-SVDD module further enhances the sensitivity of the model to small shifts in the data, reduces the false alarm rates. This combination makes the model very effective for detecting anomalies in ECG signals. Furthermore, Shah et al. [31] proposed a hybrid model, ECG-TransCovNet, which combines transformers and CNNs for superior temporal and spatial feature extraction, achieving very high performance in arrhythmia detection.

These studies demonstrate significant advancements in applying DL models to ECG monitoring. The strengths of transformer models and VAEs, such as their ability to handle complex data and generate high-quality representations, make them ideal candidates for further developments. Our proposed model builds on these advancements with the integration of Explainable AI techniques to enhance transparency and trustworthiness in ECG anomaly detection.

2.2   Transformer

The transformer network, initially presented by Vaswani et al. [40], has brought a revolutionary change in natural language processing and other fields due to its powerful capabilities in handling sequential data. Unlike recurrent neural networks (RNNs), which rely on sequential information processing through recurrent connections, transformers utilize self-attention techniques that enable them to recognize the dependencies throughout the whole sequence, regardless of the distance between elements. This allows transformers to process data more efficiently and effectively. To retain information about the position, positional encoding is incorporated into input embeddings, enabling the model to recognize the sequence order, which the architecture does not inherently encode. The attention scores (Att) of each component of the input are calculated using the following formula:

$$\mathrm{Att}(\mathbf{Q},\ \mathbf{K},\ \mathbf{V}) = \mathrm{Softmax}\Big(\frac{\mathbf{Q}\mathbf{K}^{T}}{\sqrt{d_k}}\Big)\mathbf{V} \tag{1}$$

In this formula, matrices $\mathbf{K}$ (key), $\mathbf{Q}$ (query), and $\mathbf{V}$ (value) are obtained from the input, d_k is the key vector's dimension. Instead of using a single attention mechanism, transformers employ multiple attention heads. Each attention head operates on distinct projections of the matrices $\mathbf{K},\ \mathbf{Q}$, and $\mathbf{V}$, allowing the model to recognize various patterns and correlations within the input data. These attention heads' outputs are then combined and subsequently linear transformed, resulting in the multi-head attention layer's final output.

The revolutionary multi-head attention mechanism in transformers allows for parallel processing of sequences. This parallel processing capability enables

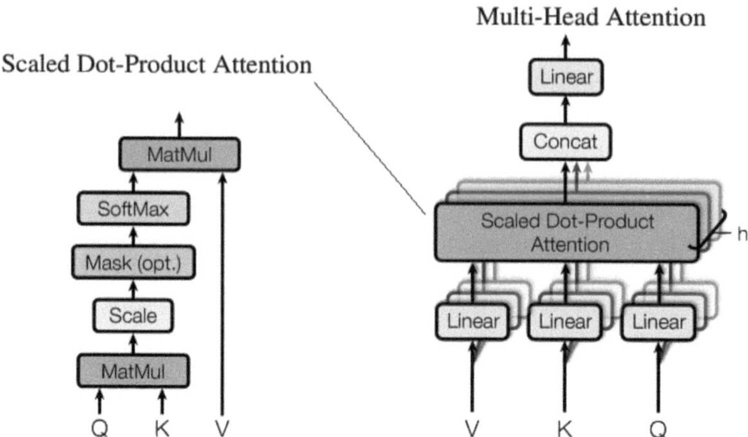

Fig. 1. Multi-Head Attention, consisting of h parallel attention layers, as introduced by Vaswani et al. [40] (Right)

transformers to understand complex data relationships, making them incredibly effective for tasks like ECG anomaly detection, where temporal and spatial dependencies are critical. Figure 1 shows how multiple attention layers run in parallel in multi-head attention introduced by Vaswani et al. [40]. Furthermore, transformers incorporate feedforward neural network layers, normalization layers, and residual connections, enhancing the model's efficiency in learning and processing sequential data. This architecture has served as the base for numerous advanced models, such as Gemini, GPT, BERT, and others, excelling in multiple NLP tasks and extending their success to other domains, as noted by Chen et al. [7]; Singh and Mahmood [34], and Acheampong et al. [1].

2.3   Variational Autoencoder (VAE)

VAEs play a crucial role in unsupervised ML, representing a significant advancement DL. Firstly introduced by Kingma et al. [13], VAEs have been widely applied in representation learning, data generation, and anomaly detection. Different from traditional autoencoders, VAEs aim to learn a probability distribution within the latent space, typically approximating a normal distribution to capture data's underlying structure.

Let X represent the input data, which is assumed to originate from unobservable latent variable Z in latent space. In VAEs, the focus is on probabilistic encoders and decoders rather than deterministic ones. Based on Bayes' theorem, the connection between $p_\theta(Z)$, $p_\theta(Z|X)$ and $p_\theta(X|Z)$, is expressed as:

$$p_\theta(Z|X) = \frac{p_\theta(X|Z) \cdot p_\theta(Z)}{\int p_\theta(X|Z) \cdot p_\theta(Z)\, dz} \tag{2}$$

In a real-world application, calculating $p_\theta(Z|X)$ directly is often intractable due to the denominator's integral. To address this challenge, Kingma et al. [13] proposed to approximate $p_\theta(Z|X)$ with a more tractable distribution $q_\phi(Z|X)$ by using variational inference, where ϕ denotes the approximate distribution's parameters. Let $KL(q_\phi(Z|X)\|p_\theta(Z|X))$ denote the Kullback-Leibler (KL) divergence, which measures the difference between two distributions:

$$KL(q_\phi(Z|X)\|p_\theta(Z|X)) = \mathbb{E}_{q_\phi(Z|X)}\left[\log\frac{q_\phi(Z|X)}{p_\theta(Z|X)}\right]. \tag{3}$$

By minimizing this divergence, the similarity of $q_\phi(Z|X)$ and $p_\theta(Z|X)$ is ensured.

Once input data is mapped to a probabilistic distribution by the encoder in the latent space, the decoder samples Z to produce data that resembles input and this sample is generally assumed to follow $\mathcal{N}(\boldsymbol{\mu}, \boldsymbol{\Sigma} = \boldsymbol{\sigma}\boldsymbol{\sigma}^T)$, i.e., $Z = \boldsymbol{\mu} + \boldsymbol{\sigma} \odot \boldsymbol{\epsilon}$. Here, $\boldsymbol{\epsilon}$ follows $\mathcal{N}(\mathbf{0}, \mathbf{I})$ and $\odot$ denotes element-wise multiplication. By incorporating $\boldsymbol{\epsilon}$, the sampling process remains differentiable. This re-parameterization trick technique is crucial for enabling the training and optimization of VAEs, making them effective for various tasks relating to representation learning and generative. The common structure of a VAE is illustrated in Fig. 2.

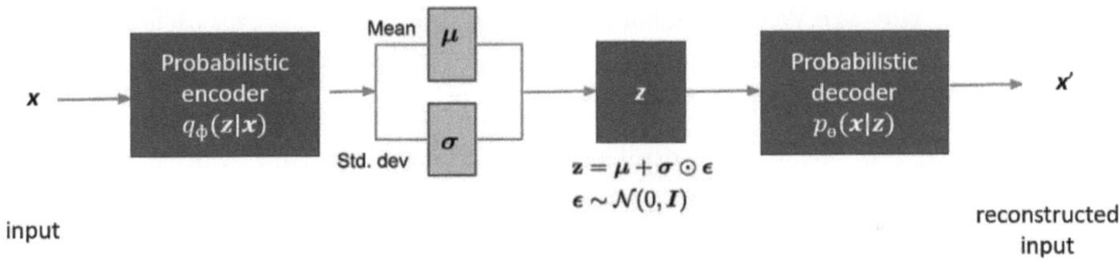

Fig. 2. VAE general structure.

The loss function of VAEs includes 2 main components: the loss from reconstruction $L(X, X')$, where X' denotes reconstructed data, and the KL divergence $KL(q_{\phi,i}(Z|X)\|p_\phi(Z))$. The reconstruction loss evaluates the ability of the decoder to reconstruct input data while the KL divergence accesses the difference between $q_\phi(Z|X)$ and $p_\phi(Z)$. The loss function of VAEs is a key component in training these models, and the goal is to minimize this function during training:

$$\text{TotalLoss (VAE)} = L(X, X') + \sum_i KL(q_{\phi,i}(Z|X)\|p_\phi(Z)). \tag{4}$$

The type of data being processed will decide the choice of reconstruction loss $L(X, X')$. Training VAEs requires minimizing this total loss function, typically via a gradient-based method, to enhance the accuracy of the reconstruction and the alignment between targeted distribution and latent space. VAEs have demonstrated their utility across various domains, such as data compression, generative modeling, and anomaly detection, as evidenced by works such as Anstine and Isayev [5], Zhang et al. [41], and Raza et al. [27].

2.4   SVDD Model Using the MEWMA Technique for Anomaly Detection

SVDD is a highly recognized algorithm in ML, especially in anomaly detection and outlier identification in datasets. First introduced by Tax and Duin [38], SVDD has become well-known for its effectiveness in various applications. One of its key strengths is its capacity to function effectively without strict assumptions regarding the data's underlying distribution, thus enhancing its versatility and robustness, as shown in many studies (Zhang and Deng [42]; Liu et al. [16]; Frusque et al. [9]). This section briefly overviews the principles of SVDD combined with the MEWMA technique, a simple yet effective method for detecting anomalies in time series data. The MEWMA approach calculates the weighted average of components in the data, utilizing a memory-based advantage to preprocess the data for utilization in the SVDD model.

SVDD is an innovative algorithm for one-class classification that excels at detecting abnormal instances by effectively modeling the normal. It establishes a sphere around the normal data, with the center's sphere denoted by a and the sphere's radius denoted by R. The main goal is to minimize the sphere's volume while still encompassing a large amount of the training samples. Given a time series dataset $S = \{x_1, x_2, \ldots, x_N\}$ where $x_i \in \mathbb{R}^p$ and p representing the count of time steps, the MEWMA vector $\mathbf{w_i} = (w_{i,1}, \ldots, w_{i,t}, \ldots, w_{i,p})$ is computed as:

$$w_{i,t} = rx_{i,t} + (1 - r)w_{i,t-1}, \tag{5}$$

where $\mathbf{w}_0 = \mathbf{0}$, r is the fixed smoothing parameter, $r \in (0, 1]$ and $t = 1, \ldots, p$. The SVDD optimization problem, aimed at describing the dataset using these MEWMA-derived vectors, is formulated as follows:

$$\min_{R,\mathbf{a}} \quad R^2 + C \sum_{i=1}^{N} \xi_i \tag{6}$$

$$\text{st.} \quad \|\mathbf{w}_i - \mathbf{a}\|^2 \leq R^2 + \xi_i, \quad i = 1, \ldots, N,$$

$$\xi_i \geq 0$$

where, C is a trade-off that helps to balance the sphere's volume and the errors. By introducing Lagrange multipliers α_i and γ_i to problem 6, we obtain the dual problem:

$$\max_{\alpha} \quad \sum_{i=1}^{N} \alpha_i \langle \mathbf{w}_i, \mathbf{w}_i \rangle - \sum_{i,j=1}^{N} \alpha_i \alpha_j \langle \mathbf{w}_i, \mathbf{w}_j \rangle \tag{7}$$

$$\text{st.} \quad \sum_{i=1}^{N} \alpha_i = 1,$$

$$0 \leq \alpha_i \leq C \quad \text{for} \quad i = 1, \ldots, N.$$

This convex quadratic problem is easier to solve compared to the primal one. Upon solving this, we obtain:

$$\mathbf{a} = \sum_{s \in SV} \alpha_s \mathbf{w}_s \tag{8}$$

$$R^2 = \langle \mathbf{w}_k, \mathbf{w}_k \rangle - 2 \sum_{i=1}^{N} \alpha_i \langle \mathbf{w}_i, \mathbf{w}_k \rangle + \sum_{i,j=1}^{N} \alpha_i \alpha_j \langle \mathbf{w}_i, \mathbf{w}_j \rangle \tag{9}$$

here, SV is the support vectors index set and for any support vector $\mathbf{w}_k$ with $0 < \alpha_k < C$, the equality in Eq. 9 holds, meaning that $\mathbf{w}_k$ lies exactly on the boundary of the hypersphere.

To monitor new samples $\mathbf{z}_1, \mathbf{z}_2, \ldots$, we convert them into MEWMA vectors $\hat{\mathbf{z}}_1, \hat{\mathbf{z}}_2, \ldots$ and determine their distance to the sphere's center $\mathbf{a}$. The squared distance is calculated as follows:

$$\|\hat{\mathbf{z}}_t - \mathbf{a}\|^2 = \langle \hat{\mathbf{z}}_t, \hat{\mathbf{z}}_t \rangle - 2 \sum_{i=1}^{N} \alpha_i \langle \mathbf{w}_i, \hat{\mathbf{z}}_t \rangle + \sum_{i,j=1}^{N} \alpha_i \alpha_j \langle \mathbf{w}_i, \mathbf{w}_j \rangle \tag{10}$$

In practical applications, kernel functions are often used instead of the inner product to achieve a more adaptable representation of data. Well-known kernel functions, such as linear, Laplacian, and Gaussian kernels, are frequently employed (Vapnik [39]; Smola et al. [35]). When using kernel function K, the squared distance is calculated as follows:

$$\mathrm{D}(\hat{\mathbf{z}}_t, \mathbf{a}) = \mathrm{D}(\hat{\mathbf{z}}_t) = \mathrm{K}(\hat{\mathbf{z}}_t, \hat{\mathbf{z}}_t) - 2 \sum_{i=1}^{N} \alpha_i \, \mathrm{K}(\mathbf{w}_i, \hat{\mathbf{z}}_t) + \sum_{i,j=1}^{N} \alpha_i \alpha_j \, \mathrm{K}(\mathbf{w}_i, \mathbf{w}_j). \tag{11}$$

2.5 SHapley Addictive ExPlanations (SHAP) - Technique

As mentioned before, among the various XAI strategies in the literature, SHAP is one of the most well-known, as proposed by Lundberg and Lee [20]. SHAP is a method based on game theory that can locally or globally interpret any machine learning or deep learning model output. This technique assigns each feature an importance value, known as Shapley scores/values, which indicate the impact of a feature over every possible combination of features. These values play a crucial role in assessing the degree of impact each feature has on the predicted outcomes of the model.

According to Lundberg and Lee [20], a method to make a complex model f more comprehensible is to approximate it with a more understandable function g. This function g is often used in explanation models, where it operates on simplified input variables z' that map to the original instances x via function h_x that ensures $x = h_x(z')$. The construction of function g is as follows:

$$g(z') = \phi_0 + \sum_{k=1}^{M} \phi_k z'_k \tag{12}$$

where $z' \in \{0,1\}^M$ a coalition of simplified input feature values z_k, where a feature's inclusion in the coalition is indicated by $z_k = 1$ (or 0, respectively), M represents the total count of simplified features, ϕ_k represents the Shapley scores of the k-th feature. These Shapley values are calculated as follows:

$$\phi_k = \sum_{S \subseteq N \setminus \{k\}} \frac{|S|!(|N| - |S| - 1)!}{|N|!} \left[v(S \cup \{k\}) - v(S) \right] \tag{13}$$

here, N represents the entire features set, $S \subset N$ excluding k, and $v(S)$ consists of prediction value of features in S. The value ϕ_k indicates the impact of feature k on the difference in the prediction of the model when feature k is included versus when it is excluded. The factorial terms $\frac{|S|!(|N|-|S|-1)!}{|N|!}$ ensure that all possible permutations of features are considered, providing a weighted average that reflects the marginal impact of each feature across every potential combination.

Many implementations of SHAP have been customized for different model types and use cases. Kernel SHAP employs an approach involving a special weighted local linear regression to compute Shapley values and is suitable for any type of ML model, although it may require significant computation (Lundberg and Lee [20]). On the other hand, tree SHAP is tailored for tree-based models like gradient-boosting machines, decision trees, random forests, etc., leveraging their structure to efficiently calculate Shapley values (Lundberg et al. [19]). Linear SHAP specifically addresses linear models, computing Shapley values by leveraging the model's linear structure for efficiency (Lundberg and Lee [20]). Deep SHAP is designed for DL models, combining SHAP values with the DeepLIFT algorithm to provide fast approximations of Shapley values for neural network models (Lundberg and Lee [20]; Shrikumar et al. [33]). Gradient SHAP, another variant, integrates SHAP values with gradients, approximating expected values by randomly sampling from a distribution defined by the input and a baseline, making it particularly effective for deep learning models (Lundberg and Lee [20]). These implementations, especially Deep SHAP and Gradient SHAP, are crucial for interpreting complex neural networks, where traditional methods may be inadequate due to the complexity of the model and non-linear properties.

SHAP is widely used in many research applications due to its versatility and effectiveness in explaining model predictions. Its applications span across various domains, including healthcare, where it has been used to interpret complex models for diagnosing diseases (Lundberg et al. [18,21]); finance, where it is broadly applied to various financial applications to improve model interpretability and transparency (Martins et al. [22]); and industrial control systems, where it improves AI-based anomaly detection methods (Do et al. [8]). The robustness and interpretability of SHAP make it a valuable tool for ensuring transparency and trust in AI systems, which is crucial for critical decision-making processes across different fields.

3 Proposed Explainable Transformer-Based Anomaly Detection Framework

This section introduces the proposed explainable framework for detecting anomalies in time series data. We will integrate the SHAP method into the existing framework, which combines a transformer-based VAE with the MEWMA-SVDD control chart (Nguyen et al. [24]). By including SHAP, we aim to make the anomaly detection process more transparent and the decisions of the model easier to understand. The combination of the transformer-based VAE and MEWMA-SVDD control chart has proven effective in managing false alarm rates, which is essential for practical applications, especially in healthcare. This integration aims to build a reliable and transparent framework that can be efficiently applied in healthcare settings, ensuring both accuracy and interpretability.

3.1 A Brief Review on Transformer-Based VAE Architecture

We utilized the transformer-based VAE framework in the study of Raza et al. [27] to derive reconstruction error vectors, which serve as the input for the MEWMA-SVDD chart. This network has shown an enhanced ability to identify abnormal signals in ECG datasets relative to other DL models.

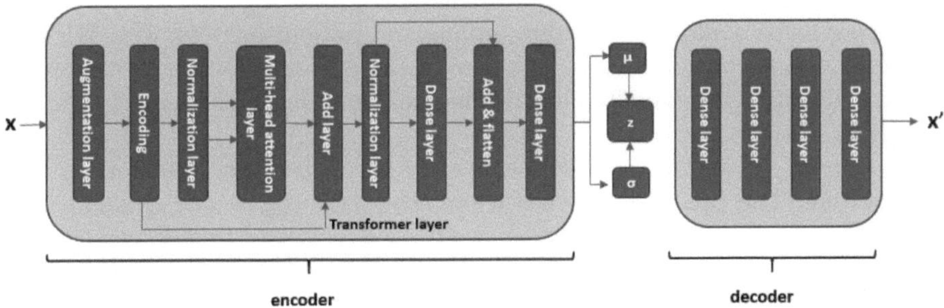

Fig. 3. Transformer-based VAE introduced in study of Raza et al. [27].

The transformer-based VAE architecture is illustrated in Fig. 3. The encoding process starts by an input layer that receives time series data. This data passes into the transformer layer, which includes several sub-layers. Subsequently, the data undergoes normalization in a normalization layer and moves to a multi-head attention layer. The weighted averages of input vectors will be calculated by this mechanism to generate the corresponding output vectors. Specifically, for a set of k-dimensional input vectors $\mathbf{x}_1, \mathbf{x}_2, \ldots, \mathbf{x}_n$, their associated output vectors $\mathbf{x}'_1, \mathbf{x}'_2, \ldots, \mathbf{x}'_n$ produced by the self-attention approach are given by:

$$\mathbf{x}'_i = \sum_j w_{i,j} \mathbf{x}_j \tag{14}$$

where the sum is calculated across the entire sequence, and $w_{i,j}$ are determined as follows:

$$w_{i,j} = \frac{\exp(\mathbf{x}_i^T \mathbf{x}_j)}{\sum_j \exp(\mathbf{x}_i^T \mathbf{x}_j)}. \tag{15}$$

By applying the softmax function, we ensure that the weights across the sequence add up to one. The decoder consists of 4 dense layers. The last layer uses a sigmoid activation to create probability distributions for potential classes. The detail of this model can be found in the study of Raza et al. [27]. This transformer-based VAE architecture, with its advanced feature extraction capabilities and robust handling of complex time-series data, significantly enhances anomaly detection accuracy, making it particularly effective for critical healthcare applications.

3.2   A Brief Review on MEWMA-SVDD Module

To effectively identify between "normal" and "abnormal" instances without relying on data distribution and to minimize false alarm rate and heighten sensitivity to subtle changes in data, Nguyen et al. [24] proposed incorporating the MEWMA-SVDD chart to the transformer-based VAE architecture mentioned in Subsect. 3.1. Reducing false alarms is vital in healthcare to ensure the safety of patients and precise diagnostics. For instance, reducing false alarms in ECG monitoring helps avoid undue interventions and patient anxiety. The MEWMA technique, which considers every instance in the current monitoring time point, enhances sensitivity to small shifts or changes, making it particularly useful in healthcare settings.

Let $\mathbf{e}_i = \mathbf{x}_i - \mathbf{x}_i'$ be the reconstruction error vector from the transformer-based VAE module. The MEWMA vector $\mathbf{w}_i = (w_{i,1}, \ldots, w_{i,t}, \ldots, w_{i,p})$ is calculated using:

$$w_{i,t} = re_{i,t} + (1 - r)e_{i,t-1}, \quad t = 1, 2, \ldots, p \tag{16}$$

here, p is the number of time steps and $\mathbf{e}_i \in \mathbb{R}^p$. The MEWMA-SVDD model is fitted using $\mathbf{w}_i$ for $i = 1, \ldots, N$, where N is the training set size. The threshold h, determined by a pre-specified ARL_0, is used as the upper limit threshold of the control chart to effectively control the false alarms. An instance $\mathbf{z}$ is classified as an abnormal if its distance $\mathrm{D}(\mathbf{z}, \mathbf{a})$ from the center $\mathbf{a}$ exceeds h, where $\mathrm{D}(\mathbf{z}, \mathbf{a})$ is calculated as in Eq. 11. The threshold h of the MEWMA-SVDD chart can be obtained using bootstrap aggregating methods as described by Tang et al. [37]. Given a training set $S = \mathbf{w}_1, \mathbf{w}_2, \ldots, \mathbf{w}_N \subset \mathbb{R}^n$, the procedure for determining h is given in Algorithm 1.

Algorithm 1: Determining the limit threshold h to manage false alarm rate.
Input: Training data set $S = \{\mathbf{w}_1, \mathbf{w}_2, \ldots, \mathbf{w}_N\}$
Output: Threshold h for MEWMA-SVDD chart

1. Generating M bootstrap samples from S.

2. For each jth bootstrap sample, where $j = 1, \ldots, M$, **do**

- Obtain the SVDD sphere by applying the SVDD algorithm in Subsection 3.2.

- Use Eq. 11 to calculate statistics $D_{j1} = \mathrm{D}(\mathbf{a}^{(j)}, \mathbf{w}_1^{(j)})$, $D_{j2} = \mathrm{D}(\mathbf{a}^{(j)}, \mathbf{w}_2^{(j)})$, $\ldots, D_{jn} = \mathrm{D}(\mathbf{a}^{(j)}, \mathbf{w}_n^{(j)})$.

- Let $\beta = \frac{1}{\mathrm{ARL}_0}$ be the chosen false alarm. Calculate $100 \times (1 - \beta)th$ percentile of n statistics $D_{jk}, k = 1, \ldots, n$, denote this value by $h^{(j)}$.

End for

3. Determine h by $100 \times (1 - \epsilon)$ percentile M values $h^{(j)}, j = 1, \ldots, M$, with ϵ is a small specified error.

This framework's performance, combining transformer-based VAE and MEWMA SVDD, has proven its effectiveness in identifying anomalies within multivariate ECG time series data (Nguyen et al. [24]). Next, we will use the XAI technique to enhance model transparency and trustworthiness. Specifically, we will incorporate the SHAP approach, as mentioned in Subsect. 2.5, into this model to provide transparent and interpretable insights into the anomaly detection process, improving decision-making and user trust.

3.3   Integration of XAI to Transformer-Based VAE Combined with MEWMA-SVDD Chart Framework

In this subsection, we detail the integration of SHAP into the framework mentioned in Subsects. 3.1, 3.2 to enhance its interpretability and transparency. SHAP is well-known for its ability to provide clear and understandable explanations for model predictions, quantifies the impact of features both globally and locally. By applying SHAP analysis globally, we assess the overall feature importance across the dataset in relation to reconstruction error, while locally, we examine individual ECG segments to understand how specific features contribute to reconstruction error in each segment. This focused application of SHAP not only deepens our understanding of how each time window influences the model's outputs but also highlights SHAP's broad applicability for providing detailed insights across all instances and for each specific instance. This approach expands the instance-specific analyses proposed by Raza et al. [27], which focuses only on specific sub-segments within individual ECG signals that contribute significantly to reconstruction loss. By providing detailed explanations at both global and individual levels, SHAP enhances the reliability and trustworthiness of our proposed anomaly framework. This makes it an invaluable tool for advancing model in critical domains like healthcare.

Figure 4 illustrates our proposed explainable framework for ECG anomaly detection. In this proposed framework, the transformer-based VAE model extracts important feature data, and the resulting reconstruction error vectors are then utilized for anomaly detection. We apply the SHAP approach to the outputs of the VAE model both globally and locally: globally, to understand the overall feature importance across the dataset, and locally, to inspect individual ECG segments for detailed insights into the reconstruction errors. This method enhances the transparency and interpretability of the anomaly detection process facilitated by our framework.

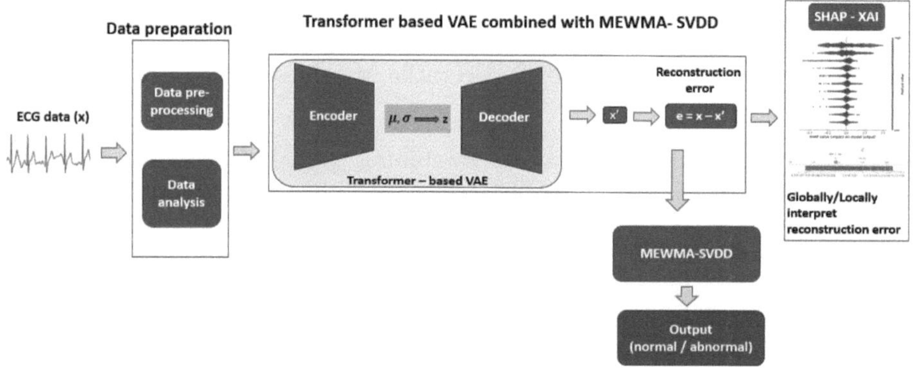

Fig. 4. The proposed explainable framework for ECG anomaly detection.

4   Experimental Procedure and Results

In this section, we first review the data utilization, experimental procedure, and the results of transformer-based VAE combined with the MEWMA-SVDD model proposed in [24]. Then, we integrate SHAP to enhance interpretability within the framework.

4.1   Brief Review of Data Utilization, Experimental Framework

In this subsection, we review the experimental results from the previously published work (Nguyen et al. [24]), which tested the effectiveness of the transformer-based VAE combined with MEWMA-SVDD chart.

Data Description. In this experiment, we combined two datasets from PhysioNet to create a hybrid dataset. The abnormal data were taken from the BIDMC Congestive Heart Failure Database, which contains ECG records from 15 patients diagnosed with severe congestive heart failure. The normal data came from the Massachusetts Institute of Technology-Beth Israel Hospital (MIT-BIH) dataset, which includes records from 18 subjects without notable arrhythmias. Each heartbeat in these datasets was pre-labelled by the expert annotators, with the labels indicating the presence of specific arrhythmias or normal rhythms. The dataset consisting of 5,000 heartbeats, including 2,919 normal and 2,081 abnormal for training and testing, was randomly selected. Figure 5 displays visual samples from these databases, with the blue line representing a normal signal and the orange one indicating an abnormal one.

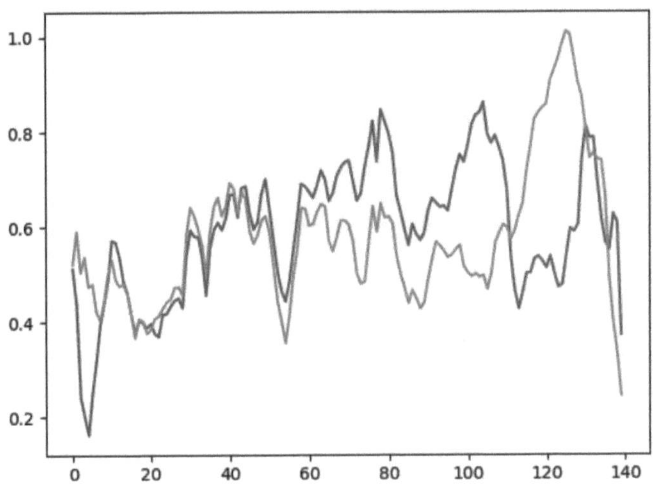

Fig. 5. An example of normal and abnormal instances from ECG dataset.

Experiment Setting. The transformer-based VAE was trained using a dataset of 4,000 randomly selected samples, reserving the remaining 1,000 samples for testing. The training process uses mean squared error to be the reconstruction loss $L(X, X')$. For the MEWMA-SVDD, the optimal bandwidth was identified through a grid search and 30-fold cross-validation. The MEWMA statistic's smoothing parameter was fixed to 0.2, and the in-control ARL ARL_0 was fixed at 100. The limit thresholds of the model were established using Monte Carlo simulations with 10,000 iterations.

Results. The effectiveness of the model was assessed using precision, recall, F1-score, and accuracy. The model obtained an impressive accuracy of 98.77%, outperforming previous approaches. The confusion matrix in Fig. 6 and the classification report in Table 1 highlight the model's effectiveness, showing a high rate of correct classifications and a significant reduction in false positives rate.

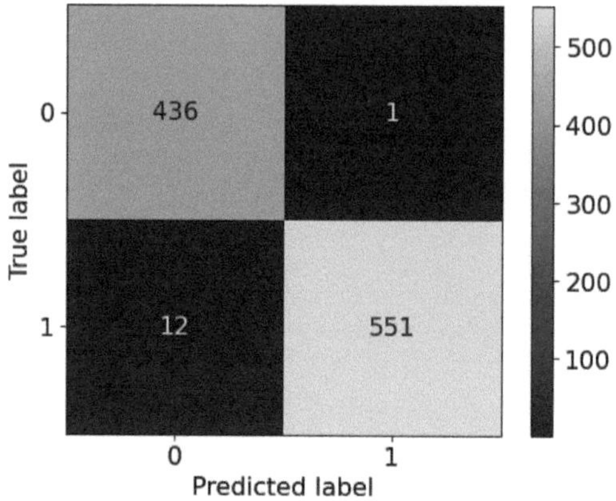

Fig. 6. Confusion matrix of the experiment.

Table 1. Classification results.

Class	Precision	Recall	F1-Score	Support
Normal	0.9977	0.9733	0.9853	437
Anomaly	0.9982	0.9787	0.9884	563
Accuracy	0.987			

Figure 7 illustrates the separation between abnormal and normal classes achieved by this framework. The test samples that lie above the threshold h are identified as anomalies, while those within the limit are identified as normal, demonstrating the model's efficiency. The code for this experiment is available at Github: Transformer-based combined with MEWMA-SVDD chart for ECG monitoring.

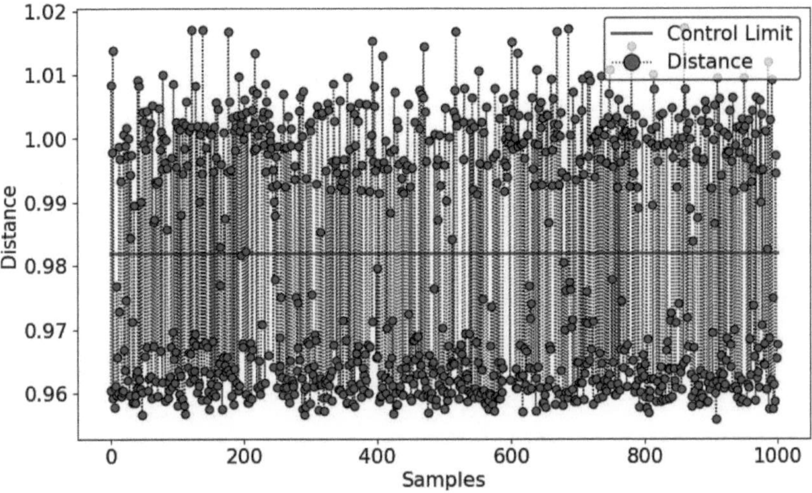

Fig. 7. Separation between normal and abnormal classes in ECG dataset.

4.2 Integration of SHAP Techniques

To enhance the proposed model's interpretability on EGG monitoring, we divided each ECG segment's 140-time steps into 14 non-overlapping windows, each containing 10-time steps. By using this windowing and averaging approach, each input feature to SHAP now represents the average reconstruction error over a window of 10 time steps rather than individual time steps. This helps in understanding which sub-segments of the ECG contribute most significantly to the model's output, simplifying the interpretation and potentially highlighting more meaningful patterns in how errors develop across the ECG sequence.

The SHAP summary plot in Fig. 8 shows the influence of each windowed segment on the reconstruction error. In the plot, each horizontal line represents a different windowed segment of the ECG, identified by labels such as "W_1-10" up to "W_131-140", with the arrangement indicating the hierarchy of their importance to the model's predictions. In this plot, the positive SHAP values indicate that the corresponding window increases the reconstruction error, contributing to the detection of an anomaly. Conversely, negative SHAP values indicate that the corresponding window decreases the reconstruction error, suggesting that the segment is less likely to be anomalous. The color coding reflects the actual values of the features, with blue representing low values and pink representing high values. This suggests that higher values within critical windows (more pink dots on the positive side of the zero line) correlate with having a greater impact on the model's prediction of anomalies. From the figure, we can see

- The plot is arranged with the most significant features at the top of the figure, indicating that "W_131-140", "W_21-30", and "W_121-130" are among the

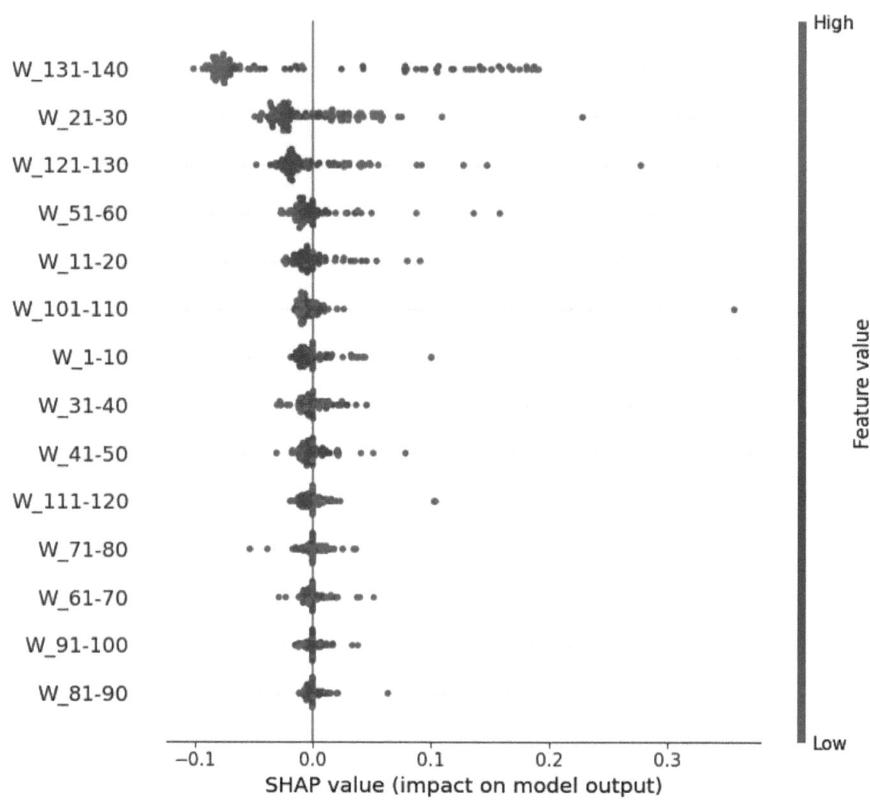

Fig. 8. SHAP plot for explaining VAE reconstruction errors.

most significant in impacting the model's output. This prioritization suggests that these windows likely contain key ECG data patterns critical for the model's predictions. For clinicians, focusing on these key segments could aid in identifying significant deviations from normal ECG patterns, which might correspond to cardiac events requiring immediate attention.

– Conversely, the features positioned at the bottom of the plot, such as W_81-90", W_91-100", and W_61-70", have less impact on the model's decision-making process. Variations within these segments, whether high or low, do not significantly influence the model's predictions of anomalies. For treatment and monitoring, anomalies detected in these lower-impact segments might require a different level of clinical urgency than those identified in the top segments. Understanding this distinction can help clinicians prioritize patient monitoring and intervention strategies based on the likelihood and potential severity of cardiac anomalies detected.

Thus, this global SHAP plot provides a comprehensive overview of the dataset, highlighting which ECG segments commonly influence anomaly detection. Clinicians can leverage these global insights to improve diagnostic precision. By focusing on the most influential segments, they can efficiently screen for anomalies, enabling fast and precise evaluations in various medical settings. This plot's transparency about how data influences predictions boosts trust in the model. It enables clinicians to identify which ECG patterns are considered important, matching AI results with well-known medical evidence. This alignment confirms that the model's predictions are reliable and based on established practices in heart care.

Besides the global interpretation provided by the SHAP summary plot, a detailed local understanding can be gained by examining individual predictions using a SHAP force plot. This type of visualization specifically highlights how each ECG windowed segment contributes to the model's prediction for a particular case, offering precise insights into the factors contributing to anomalies in each individual instance. Figure 9 shows the force plot for an individual abnormal instance from the ECG5000 dataset and represents the impact of particular windows on the reconstruction errors. Red segments indicate that the corresponding feature increased the model's predicted reconstruction error relative to the base value, suggesting that these windows are more difficult for the VAE model to reconstruct accurately. This difficulty may be associated with anomalies or irregularities in these segments. Blue segments, on the other hand, suggest that the presence of these windows in the input data decreases the overall error, potentially indicating segments that are easier to reconstruct accurately, likely because they align more closely with the normal patterns. From the plot, we can see that window segments "$W\_131\text{-}140$", "$W\_21\text{-}30$" and "$W\_101\text{-}110$" exhibit higher positive SHAP values, which significantly and positively impact the model's output. This suggests that these segments may contain potential sub-segments that define the abnormal pattern in an ECG. Conversely, windows such as "$W\_121\text{-}130$" or "$W\_51\text{-}60$" and "$W\_11\text{-}20$ contribute negatively (lowering the error) to the reconstruction error, indicating areas where the model performs well, thereby suggesting these segments align more closely with the normal patterns of ECG. For a more detailed visualization, these SHAP values are plot together with the real ECG segment, as shown in Fig. 10. The top three contributing windows with the largest SHAP values, which are likely to contain abnormal patterns, are highlighted in varying intensities of red color. These highlighted segments may correspond to the region where VAE model struggles the most in reconstructing ECG signal.

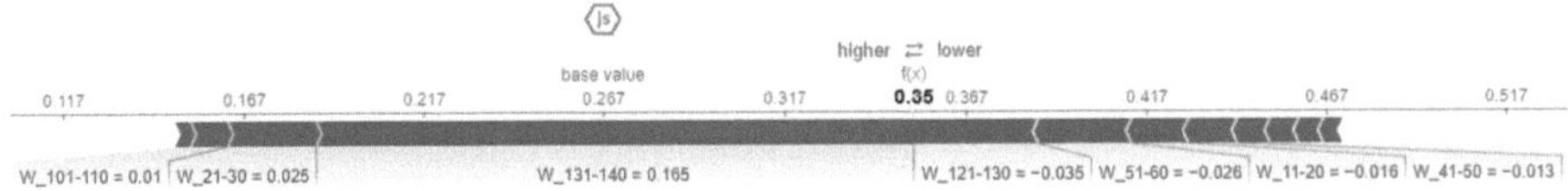

Fig. 9. SHAP force plot for explaining VAE reconstruction error in a specific instance.

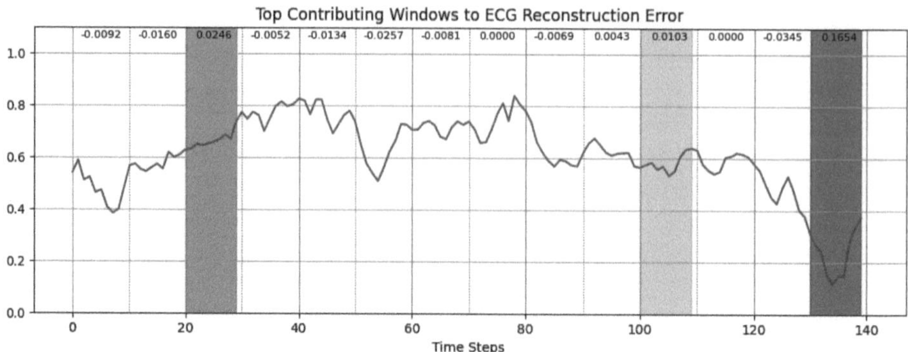

Fig. 10. SHAP Values with highlighted windows contributing to VAE reconstruction error.

To further illustration, another ECG segment is analyzed and visualized in Figs. 11 and 12. According to these figures, the window segments "$W\_131\text{-}140$", "$W\_121\text{-}130$" and "$W\_31\text{-}40$" are highlighted and may indicate the critical patterns that are potential indicator of abnormal cardiac activities. Clinicians should firstly focus on these highlighted segment during their review before further diagnostic testing.

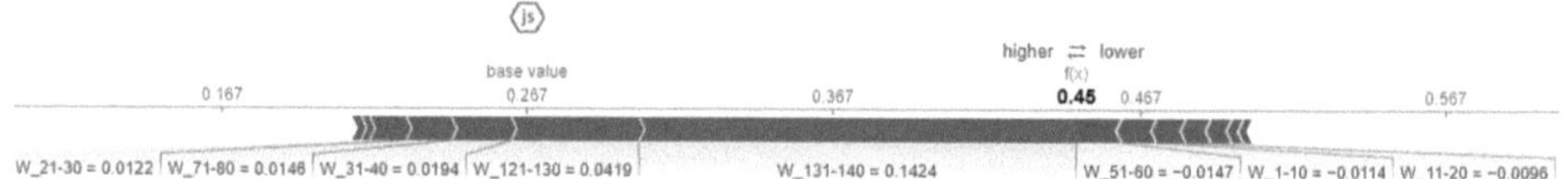

Fig. 11. SHAP force plot for explaining VAE reconstruction error in a specific instance.

Overall, the SHAP value reveals the specific windows of time steps within the ECG segments that are more influential in contributing to the reconstruction errors identified by the VAE model, both at a global and local level. By focusing on these influential windows, we can better understand which segments of the ECG signal are most contributing to anomalies, thereby improving the model's transparency and aiding clinical interpretation.

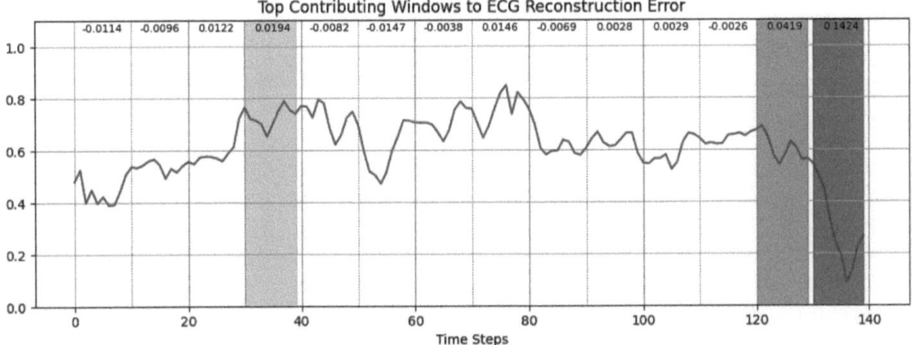

Fig. 12. SHAP Values with highlighted windows contributing to VAE reconstruction error.

5   Conclusion

In this work, we introduced an explainable framework for ECG anomaly detection by integrating XAI into a transformer-based VAE architecture combined with the MEWMA-SVDD chart. Our approach emphasizes the critical need for transparency and interpretability in AI-based models, particularly in sensitive sectors such as healthcare. The transformer-based VAE effectively extracts features from complex ECG time-series data, while the MEWMA-SVDD control chart improves anomaly detection accuracy and reduces false alarm rates. By incorporating SHAP into this framework, we provide detailed, feature-level insights to the contributions of window time steps to the model's predictions. Our applications of SHAP at both global and local levels demonstrate its robustness in explaining the model's decision-making process. This dual application highlights the most influential window time steps in the ECG data and clarifies how these windows impact anomaly classification, offering comprehensive transparency. The detailed explanations of the experiment on the ECG dataset provided by SHAP enhance the model's usability and reliability, making it a valuable tool for healthcare professionals.

Our future works will focus on integrating our framework into federated learning environments to enhance data privacy, employing self-supervised learning techniques to improve model performance with minimal labeled data, and developing XAI-by-design models to further enhance explainability and robustness in AI systems.

Acknowledgments. The authors gratefully acknowledge HEC Liège, Management School, University of Liège, Belgium and GEMTEX, ENSAIT, University of Lille, France for providing the infrastructure and resources essential to our research. We also thank the University of Liège for financially supporting the PhD program through research grants.

Disclosure of Interests. The authors declare that they have no competing interests.

References

1. Acheampong, F.A., Nunoo-Mensah, H., Chen, W.: Transformer models for text-based emotion detection: a review of BERT-based approaches. Artif. Intell. Rev. 1–41 (2021)
2. Alabi, R.O., Elmusrati, M., Leivo, I., Almangush, A., Mäkitie, A.A.: Machine learning explainability in nasopharyngeal cancer survival using lime and shap. Sci. Rep. **13**(1), 8984 (2023)
3. Albahri, A.S., et al.: A systematic review of trustworthy and explainable artificial intelligence in healthcare: assessment of quality, bias risk, and data fusion. Inf. Fusion (2023)
4. Amini, K., Mirzaei, A., Hosseini, M., Zandian, H., Azizpour, I., Haghi, Y.: Assessment of electrocardiogram interpretation competency among healthcare professionals and students of ardabil university of medical sciences: a multidisciplinary study. BMC Med. Educ. **22**(1), 448 (2022)
5. Anstine, D.M., Isayev, O.: Generative models as an emerging paradigm in the chemical sciences. J. Am. Chem. Soc. **145**(16), 8736–8750 (2023)
6. Bolhasani, H., Mohseni, M., Rahmani, A.M.: Deep learning applications for IoT in health care: a systematic review. Inform. Med. Unlocked **23**, 100550 (2021)
7. Chen, H., et al.: Pre-trained image processing transformer. In: Proceedings of the IEEE/CVF Conference on Computer Vision and Pattern Recognition, pp. 12299–12310 (2021)
8. Do, T.H., Nguyen, X.H., Nguyen, V.H., Nguyen, H.D., Truong, T.H., Kim, P.T.: Explainable anomaly detection for industrial control system cybersecurity. IFAC-PapersOnLine **55**(10), 1183–1188 (2022). 10th IFAC Conference on Manufacturing Modelling, Management and Control MIM 2022
9. Frusque, G., Mitchell, D., Blanche, J., Flynn, D., Fink, O.: Non-contact sensing for anomaly detection in wind turbine blades: a focus-SVDD with complex-valued auto-encoder approach. Mech. Syst. Signal Process. **208**, 111022 (2024)
10. Gao, H., Jia, Z.: Detection of threats under inattentional blindness and perceptual load. Curr. Psychol. **36**, 733–739 (2017)
11. Gaudilliere, P.L., Sigurthorsdottir, H., Aguet, C., Van Zaen, J., Lemay, M., Delgado-Gonzalo, R.: Generative pre-trained transformer for cardiac abnormality detection. In: 2021 Computing in Cardiology (CinC), vol. 48, pp. 1–4. IEEE (2021)
12. Jang, J.H., Kim, T.Y., Lim, H.S., Yoon, D.: Unsupervised feature learning for electrocardiogram data using the convolutional variational autoencoder. PLoS One **16**(12) (2021)
13. Kingma, D., Welling, M., et al.: An introduction to variational autoencoders. Found. Trends® Mach. Learn. **12**(4), 307–392 (2019)
14. Kreitz, C., Furley, P., Memmert, D., Simons, D.: Inattentional blindness and individual differences in cognitive abilities. PLoS ONE **10**(8) (2015)
15. Lang, M.: A low-complexity model-free approach for real-time cardiac anomaly detection based on singular spectrum analysis and nonparametric control charts. Technologies **6**(1), 26 (2018)
16. Liu, B., et al.: Adaboost-based SVDD for anomaly detection with dictionary learning. Expert Syst. Appl. **238**, 121770 (2024)
17. Liu, H., Zhao, Z., Chen, X., Yu, R., She, Q.: Using the VQ-VAE to improve the recognition of abnormalities in short-duration 12-lead electrocardiogram records. Comput. Methods Programs Biomed. **196**, 105639 (2020)

18. Lundberg, S.M., et al.: From local explanations to global understanding with explainable AI for trees. Nat. Mach. Intell. **2**(1), 56–67 (2020)
19. Lundberg, S.M., Erion, G.G., Lee, S.I.: Consistent individualized feature attribution for tree ensembles. arXiv preprint arXiv:1802.03888 (2018)
20. Lundberg, S.M., Lee, S.I.: A unified approach to interpreting model predictions. In: Advances in Neural Information Processing Systems, vol. 30, Curran Associates, Inc. (2017)
21. Lundberg, S.M., et al.: Explainable machine-learning predictions for the prevention of hypoxaemia during surgery. Nat. Biomed. Eng. **2**(10), 749–760 (2018)
22. Martins, T., De Almeida, A.M., Cardoso, E., Nunes, L.: Explainable artificial intelligence (XAI): a systematic literature review on taxonomies and applications in finance. IEEE Access (2023)
23. Muzammil, M.A., et al.: Artificial intelligence-enhanced electrocardiography for accurate diagnosis and management of cardiovascular diseases. J. Electrocardiol. (2024)
24. Nguyen, T.T.V., Heuchenne, C., Tran, K.D., Tran, K.P.: A novel transformer-based anomaly detection approach for ECG monitoring healthcare system. In: International Conference on Safety and Security in IoT, pp. 111–129. Springer, Cham (2023)
25. Panwar, H., Gupta, P., Siddiqui, M.K., Morales-Menendez, R., Bhardwaj, P., Singh, V.: A deep learning and grad-cam based color visualization approach for fast detection of covid-19 cases using chest x-ray and ct-scan images. Chaos Solitons Fractals **140**, 110190 (2020)
26. Prendin, F., Pavan, J., Cappon, G., Del Favero, S., Sparacino, G., Facchinetti, A.: The importance of interpreting machine learning models for blood glucose prediction in diabetes: an analysis using shap. Sci. Rep. **13**(1), 16865 (2023)
27. Raza, A., Tran, K.P., Koehl, L., Li, S.: Anofed: adaptive anomaly detection for digital health using transformer-based federated learning and support vector data description. Eng. Appl. Artif. Intell. **121**, 106051 (2023)
28. Ribeiro, M.T., Singh, S., Guestrin, C.: "Why should I trust you?": explaining the predictions of any classifier. In: Proceedings of the 22nd ACM SIGKDD International Conference on Knowledge Discovery and Data Mining, pp. 1135–1144 (2016)
29. Saint-Lot, J., Imbert, J., Dehais, F.: Red alert: a cognitive countermeasure to mitigate attentional tunneling. In: Proceedings of the ACM Conference on Human Factors in Computing Systems (2020)
30. Selvaraju, R.R., Cogswell, M., Das, A., Vedantam, R., Parikh, D., Batra, D.: Grad-cam: visual explanations from deep networks via gradient-based localization. In: Proceedings of the IEEE International Conference on Computer Vision, pp. 618–626 (2017)
31. Shah, H.A., Saeed, F., Diyan, M., Almujally, N.A., Kang, J.M.: ECG-transcovnet: a hybrid transformer model for accurate arrhythmia detection using electrocardiogram signals. CAAI Trans. Intell. Technol. (2024)
32. Shin, D.H., Park, R.C., Chung, K.: Decision boundary-based anomaly detection model using improved anogan from ECG data. IEEE Access **8**, 108664–108674 (2020)
33. Shrikumar, A., Greenside, P., Shcherbina, A., Kundaje, A.: Not just a black box: learning important features through propagating activation differences. arXiv preprint arXiv:1605.01713 (2016)
34. Singh, S., Mahmood, A.: The NLP cookbook: modern recipes for transformer based deep learning architectures. IEEE Access **9**, 68675–68702 (2021)

35. Smola, A., Scholkopf, B., Muller, K.: The connection between regularization operators and support vector kernels. Neural Netw. 637–649 (1998)
36. Suman, G., Prajapati, D.: Control chart applications in healthcare: a literature review. Int. J. Metrol. Qual. Eng. **9**, 5 (2018)
37. Tang, A., Castagliola, P., Hu, X., Xie, F.: An assessment for the conditional performance of an support vector data description (SVDD)-based chart. Qual. Reliab. Eng. Int. 1–17 (2022)
38. Tax, D.M.J., Duin, R.P.W.: Support vector data description. Mach. Learn. **54**(1), 45–66 (2004)
39. Vapnik, V.N.: Statistical Learning Theory. Wiley, Hoboken (1998)
40. Vaswani, A., et al.: Attention is all you need. In: Advances in Neural Information Processing Systems, vol. 30, Curran Associates, Inc. (2017)
41. Zhang, C., et al.: VESC: a new variational autoencoder based model for anomaly detection. Int. J. Mach. Learn. Cybern. **14**(3), 683–696 (2023)
42. Zhang, Z., Deng, X.: Anomaly detection using improved deep SVDD model with data structure preservation. Pattern Recogn. Lett. **148**, 1–6 (2021)
43. Zhou, H., Kan, C.: Tensor-based ECG anomaly detection toward cardiac monitoring in the internet of health things. Sensors **21**(12), 4173 (2021)

AI for Human Services and Social Applications

Educable Learning for Human-AI Coevolution

Frédéric Vanderhaegen[1,2]([⊠]) [ID]

[1] Université Polytechnique Hauts-de-France, LAMIH UMR CNRS 8201, Le Mont Houy, 59313 Valenciennes Cedex 9, France
`frederic.vanderhaegen@uphf.fr`
[2] INSA Hauts-de-France, Le Mont Houy, 59313 Valenciennes Cedex 9, France

Abstract. Human-AI coevolution supposes that knowledge of human and AI system evolves in the course of their activities. This knowledge evolution can be done by applying human-centered shared autonomy between human and AI systems. It includes the discovery and the updating of knowledge. To do so, this paper proposes a new feature for AI-based system: the educable learning process. It consists in making system educable by discovering and updating technical or human knowledge when inconsistency between knowledge is occurring in the course of use experience. Knowledge is modelled with natural human reasoning principles and with merged or cumulative machine learning processes. The detection of inconsistencies between knowledge generates two processing: knowledge discovery and knowledge updating. The former aims to link inputs with outputs of knowledge and to explore new possible links. The latter consists in updating existing knowledge by translating and transferring it via allocation and communication system to humans or AI systems. Educable learning process is illustrated by a practical use-case study in transportation domain by discovering and updating user's knowledge to a cumulative AI system.

Keywords: Shared autonomy · human-AI coevolution · inconsistency · human reasoning · merged and cumulative learning · educable learning

1 Introduction

Concepts about industry of the future highlight the role of humans in the control and supervision loop of sociotechnical systems. Future human-AI systems will be digital, connected and assistive. Assistive technology is more and more based on Artificial Intelligence (AI) that makes it smarter and more reliable. Due to this high level of technical evolution, human tasks are regularly replaced by machines and autonomous systems are emerging. However, human expertise is still necessary when AI system fails. The development of human-centered shared autonomy between humans and machines is then an important issue to make it possible for humans to recover a failing system. Such failures may arise from a lack of consistency of AI system knowledge and sharing knowledge between human and machine supposes that there is no inconsistency in the sharing processing. As a matter of fact, users of AI system have no idea about

S. Thomassey et al. (Eds.): RAIDS 2024, LNICST 673, pp. 111–123, 2026.
https://doi.org/10.1007/978-3-032-14055-5_8

the knowledge implemented on it by designers or about the evolution algorithm of its knowledge over time. After a period of time of its use, they may build a mental model of its functioning but this model can be wrong. Therefore, mutual exchange between human and AI system are needed continually to minimize such problems. To do so, the paper proposes a new feature for human-machine system: the educable learning that makes possible this mutual exchange via allocation and communication systems when inconsistencies are detected. Inconsistencies based on weak signals mean that humans or AI systems are not aware of their existence before they occur, and face the problem of having low number of data available to process them. Therefore, knowledge discovery and updating are mandatory. Educable learning concept implies that humans and AI systems must be educable by implementing abilities for knowledge discovery and updating. This paper proposes a general framework for educable learning from natural human reasoning to AI systems.

Section 2 of the paper presents related works on human-machine interactions and human role in future systems. It motivates the interest of developing educable learning processing for knowledge discovery and updating by exploring knowledge inconsistency regarding use experience. Section 3 introduces the educable learning process into human-machine coevolution and shared autonomy context. Section 4 develops principles of educable learning that is illustrated by a use-case study in the last section. Conclusion section highlights the contributions of the paper and proposes some future research works.

2   Related Works

Regarding human-machine interactions, Industry 4.0 era emphasizes the development of supports for facilitating human-AI cooperation or collaboration or for ensuring human well-being [1–3] while that for Industry 5.0 focuses on other challenging issues like human-AI symbiosis, human-AI coevolution or human resilience [4–6]. Supports for well-being monitoring aim at detecting any situations that can make human err. Their design can apply offline analysis methods [7–10] or online measurement tools like subjective evaluation techniques, eye-trackers or connected watches to assess factors like attention, workload, trust or stress [11–16]. Human-AI cooperation or collaboration is a way to recover human inattention, overload or stress and to support human well-being. However, from the viewpoint of ecological coevolution [17–20], human-AI symbiosis can relate to cooperative or competitive interactions in terms of benefits or disadvantages respectively. On cooperative coevolution, there are benefits for each decision-makers or for a part of them with no negative or positive consequences for the others. On competitive one, some decision-makers impact negatively and positively other decision-makers or some decision-makers depend on others by impacting them negatively. These positive or negative consequences can impact human, technical or organizational factors. When unexpected hazardous impacts are recovered successfully, human-AI system is resilient otherwise it is vulnerable [21–24].

In order to limit the occurrence of such situations or to facilitate their mutual recovering, the integration of AI-based tools on workplaces implies that humans are able to perceive, understand and anticipate what these tools did, are doing or will do. To do so, the design of allocation and communication systems and their associated modalities of interactions is required [25–29]. It has to facilitate human factors like situation awareness or sensemaking. Situation awareness is the ability to perceive, evaluate current situations and anticipate their evolution [30–34] while sensemaking is the ability to correlate understanding of situations with information support content [35–37]. Both concepts about situation awareness and sensemaking are considered in the design of AI-based tools in terms of explainable AI [38–41]. Explainable AI needs several abilities like transparency to make solutions obvious and traceable for humans, interpretability or understandability to make sense to humans, or explainability to answer to what happens, how does it happen and why does it happen [38, 39]. Such developments in human-machine teaming may then increase human trust in the AI-based tool's decisions [40, 41]. In the context of user-centered design, others features are developed like usability, learnability or memorability [42, 43]. The two first concepts relates to the ease with which users appropriate a system or product and the third one concerns the ease with which a user can remember how to regain control of a system after a period of time without having to relearn how it works or how to use it.

User-centered design offers then two possible ways of design: the design of perfect AI-based system that can satisfy users' needs, or the human-AI coevolution development. The first one aims to implement abilities like transparency, interpretability, understandability, explainability, learnability, usability or memorability and to define measurements to evaluate these features. The second one involves cooperative or competitive behaviors with regard to consequences they produce to each decision-maker. The idea of coevolution supposes that human and AI-support tools are able to mutually evolve and being improved by analyzing negative and positive impact of their interactions. It depends on the level of shared autonomy between humans and AI systems that can evolve from an entire manual control to an entire automated one, with intermediate levels ranging from low to high level of automation through which human involvement in control tasks is progressively reduced and limited to monitoring the operation of the system with possible intervention in case of system failures [40, 44–49]. In this takeover case, the "AI assisted by human" process can complement the "human assisted by AI" one [50–52] to design an all-inclusive human-AI system that can meet the needs of all users regardless of their physical, cognitive, cultural or social level. Such mutual contribution of humans and AI systems exists for labeling or relabeling data, giving them meaning and facilitating their use for AI-based learning process [53–55].

Mutual contribution related to human-machine coevolution can also exist in managing knowledge. It can be built from data mines [56, 57], weak signals [58, 59] or inconsistencies [59–62]. Data mining for AI systems assumes that the amount of data is high enough to be able to deal with it. When an unprecedented hazardous situation occurs, users can complete it because the associated data to be processed is weak. Weak signals are signals with low levels of amplitude or occurrence and are by nature considered insignificant whereas they can be precursors to risks. Indeed, they can relate to dangerous scenarios or inconsistencies between human or technical knowledge that may not be taken into account during the learning process of AI systems. Weak signal-based inconsistency can be generated by the presence of other decision-makers with other implemented shared autonomy levels [62, 63].

Next section proposes a configuration of human-AI coevolution and shared autonomy principles based on educable learning to discover and update knowledge structure and content in case of such weak signal-based inconsistency occurrence detected through human feedback of experience.

3   Human-AI Coevolution, Shared Autonomy and Educable Learning

In human-AI coevolution, allocation and communication system supports the human-AI interactions, the communication of human or technical intentions or of information from the controlled process, and the exchange of knowledge (see Fig. 1). It also controls the task allocation managed by AI systems or humans.

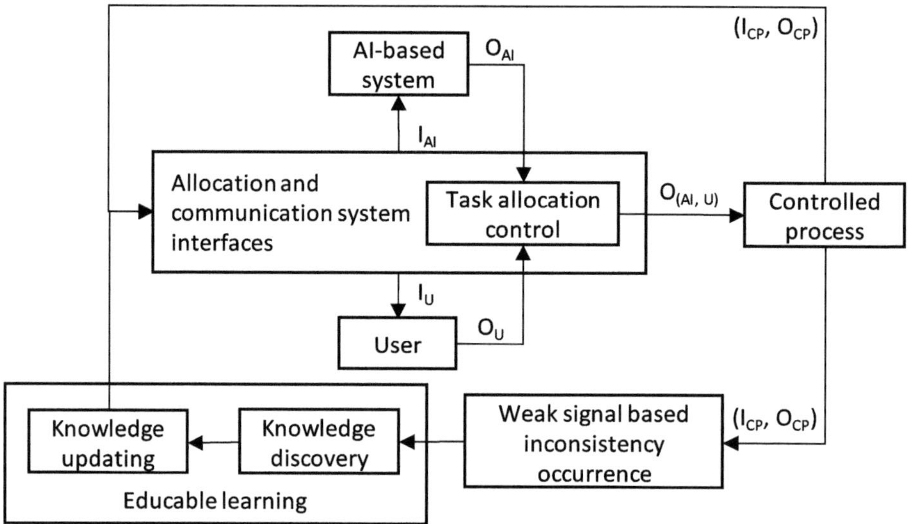

Fig. 1. Human-AI coevolution and shared autonomy for educable learning.

For each case, the emergence of inconsistency based on weak signals generates the discovery of relevant knowledge and requires the updating of the structure and content of reasoning supports. AI-based system uses knowledge base K_{AI} that contains relations between inputs I_{AI} with outputs O_{AI} and users produce a knowledge base K_U to analyze input data I_U and define an output decision O_U. The decisional process depends on input data from the controlled process I_{CP} or couples of input and output data (I_{CP}, O_{CP}) at previous control iterations. Different levels of shared autonomy can be implemented, and convergence or divergence of human or technical decision makes coevolution cooperative or competitive respectively. Such convergence or divergence can occur through different behaviors and the output $O_{(AI, U)}$ is equal to:

1. $O_{AI} \oplus O_U$: One decision-maker takes priority over the other or assumes responsibility for acting alone or thinks do better than the other. User takes over AI or vice-versa, i.e. one decides alone independently of the other.

2. $O_{AI} + O_U$: A compromise is found between the decision makers. One of them can propose a solution to the other but this solution can be applied or not, or even improve individually or mutually by the other.
3. $O_{AI} . O_U$: Decision-makers are mutually dependent on each other. Task is divided into subtasks that are achieved by user or AI system independently but the success of a subtask managed by one depends on that of the other.

The first output $O_{AI} \oplus O_U$ is for competitive coevolution while the two others for cooperative one. For both coevolution types, couples (I_{CP}, O_{CP}) can contain inconsistences that can be considered as insignificant because they relate to weak signals in the course of users' experience. This implies the genesis of knowledge discovery to validate the interest to create or modify knowledge, and of knowledge updating to do so by formatting this knowledge with respect to human and AI system reasoning supports. Knowledge discovery and updating are detailed on next section.

4   Knowledge Discovery and Updating for Educable Learning

Knowledge discovery is composed by two levels of treatment of a knowledge containing relations between inputs I_{CP} and outputs O_{CP} (see Fig. 2). Knowledge linking level builds relations between inputs and outputs and knowledge exploring level creates new ones considering weak signals-based possible inconsistencies. Knowledge discovery includes processing of merged or cumulative AI reasoning [64–66] or of natural human reasoning like deductive, abductive, inductive and counterfactual reasoning [67–74]. Elements of K_{AI} in merged and cumulative AI reasoning are neuronal models or case-based reasoning models respectively that contains vectors of data about relations between inputs and outputs. Regarding inputs I_{CP} or outputs O_{CP}, the LINK function identifies the corresponding output vector O_{CP} or input vector I_{CP} respectively that is the most similar to I_{win} or O_{win}. Input I_{CP} or output O_{CP} are then associated to I_{win} or O_{win}. Deductive and abductive reasoning interpret relation between inputs and output of K_U in terms of rules or heuristics. On deductive reasoning, when input I_U equal to I_{CP} is observed, and if a rule that makes relation between the input I_U and an output O_U noted $(I_U \rightarrow O_U)$ exists, then output O_{CP} equal to O_U can be observed. Abductive reasoning is similar but it is based on the observation of output O_{CP} equal to O_U and the existence of a rule $(I_U \rightarrow O_U)$. Abductive reasoning leads to the possibility of observing I_{CP} equal to I_U. Inductive and counterfactual reasoning aim to discover knowledge or to identify inconsistency between knowledge. For instance, the discovery of new relations between inputs and outputs can result to the production of a general rule $(I_U \rightarrow O_U)$ in the course of the inductive reasoning. On counterfactual reasoning, the discovery of possible contradictory relations between inputs and outputs with regard to field observations for instance can lead to the discovery of inconsistency between rules like $(I_{U1} \rightarrow O_U)$ and $(I_{U2} \rightarrow Not(O_U))$. This aims at producing a new rule like $(I_{U1} . I_{U2} \rightarrow O_U \oplus Not(O_U))$.

The discovery of inconsistencies or new rules involves redefining the structure and content of knowledge. This restructuring is handled by the knowledge updating process by translating and transferring human knowledge to AI systems (see Fig. 3). Knowledge of users K_U is composed by rules. Each rule noted R makes relations between inputs I and outputs O. Both inputs and outputs are composed by triplets (label, "=", value). The

---

Knowledge linking based on natural human reasoning
Deductive reasoning:
$$I_{CP}=I_U \wedge (I_U \rightarrow O_U) \in K_U \vdash O_{CP}=O_U$$
Abductive reasoning:
$$O_{CP}=O_U \wedge (I_U \rightarrow O_U) \in K_U \vdash I_{CP}=I_U$$

Knowledge linking based on merged or cumulative AI reasoning

$LINK(I_{CP}, K_{AI})$:
$$(I_{win} \cup O_{win}) \leftarrow (I_{AI1} \cup O_{AI1})$$
For all $(I_{AIp} \cup O_{AIp}) \in K_{AI}$ do
   If $(|I_{CP}^T - I_{win}^T| > |I_{CP}^T - I_{AIp}^T|)$ then
     $(I_{win} \cup O_{win}) \leftarrow (I_{AIp} \cup O_{AIp})$
   End if
End For
$O_{CP} \leftarrow O_{win}$

$LINK(O_{CP}, K_{AI})$:
$$(I_{win} \cup O_{win}) \leftarrow (I_{AI1} \cup O_{AI1})$$
For all $(I_{AIp} \cup O_{AIp}) \in K_{AI}$ do
   If $(|O_{CP}^T - O_{win}^T| > |O_{CP}^T - O_{AIp}^T|)$ then
     $(I_{win} \cup O_{win}) \leftarrow (I_{AIp} \cup O_{AIp})$
   End if
End For
$I_{CP} \leftarrow I_{win}$

Knowledge exploring based on natural human reasoning
Inductive reasoning:
$$\forall k, I_{Uk} \wedge (I_{Uk} \rightarrow O_{Uk}) \vdash K_U \leftarrow K_U \cup (I_U \rightarrow O_U)$$
Counterfactual reasoning:
$$I_{U1}, I_{U2}, I_{U1} \neq I_{U2}, (I_{U1} \rightarrow O_U) \wedge (I_{U2} \rightarrow Not(O_U)) \vdash K_U \leftarrow K_U \cup (I_{U1}.I_{U2} \rightarrow O_U \oplus Not(O_U))$$

Fig. 2. Knowledge linking and exploring principles for knowledge discovery

name of the corresponding variable to be used is "label", and the content of this variable is compared with that of "value" via the character "=" for input data of I. This character "=" aims to affect the content of "value" to the variable "label" for output data of O. A translation matrix TM is needed to translate the content of K_U into K_{AI}. It contains a list of triplets (label, value1, value2) by limiting the contents of each variable "label" to two values "value1" or "value2". The order of appearance of "label" in TM is the order of recording of data in AI systems. Therefore, the TRANSLATE function builds couples (I_m, O_m) to be transferred into AI systems by making conformity between contents of "label" of K_U with those of TM and by replacing "value1" by 0 and "value2" by 1. Then, the TRANSFER functions use these couples (I_m, O_m) to be implemented into K_{AI} of merged or cumulative AI. On merged AI process, the couple (I_{win}, O_{win}) that is the more similar to (I_m, O_m) is identified and errors ε between (I_{win}, O_{win}) with (I_m, O_m) and with the content of other neurons (I_p, O_p) is used to update K_{AI} with a given function $\Delta(\varepsilon)$. On cumulative AI process, the vector INCONSISTENCY lists same inputs I for which outputs O are opposite. If the couple (I_m, O_m) does not exists, i.e. the value of the variable EXIST is FALSE, it is integrated into K_{AI} as a new possible scenario.

Knowledge translating

$\forall\ R \in K_U,\ R=(I{\rightarrow}O),\ I=\cup\{(\text{label},\,"="\,,\text{value})\},\ O=\cup\{(\text{label},\,"="\,,\text{value})\}$
$\forall\ T \in TM,\ T=(\text{label},\text{value1},\text{value2})$

TRANSLATE$((K_U, K_{AI})$:
```
For all R=(I→O)∈K_U do
   For all i∈I do
      If ∃k, TM(k)(0)=i(0) then
         If TM(k)(1) = i(2) then
            I_m(k)=0
         Else
            I_m(k)=1
         End If
      End If
   End For
   For all o∈O do
      If ∃k, TM(k)(0)=o(0) then
         If TM(k)(1) = o(2) then
            O_m(k)=0
         Else
            O_m(k)=1
         End If
      End If
   End For
End For
```

Knowledge transferring for merged AI reasoning

TRANSFER$((I_m, O_m), K_{AI})$:
$(I_{win} \cup O_{win}) \leftarrow (I_1 \cup O_1)$
For all $(I_p \cup O_p) \in K_{AI}$ do
   If $(|(I_m \cup O_m)^T - (I_{win} \cup O_{win})^T| > |(I_m \cup O_m)^T - (I_p \cup O_p)^T|)$ then
      $(I_{win} \cup O_{win}) \leftarrow (I_p \cup O_p)$
   End if
End For

$\varepsilon^T \leftarrow |(I_m \cup O_m)^T - (I_{win} \cup O_{win})^T|$
$(I_{win} \cup O_{win})^T \leftarrow (I_{win} \cup O_{win})^T + \Delta(\varepsilon^T)$
For all $(I_p \cup O_p) \in K_{AI},\ (I_p \cup O_p) \neq (I_{win} \cup O_{win})$ do
   $\varepsilon^T \leftarrow |(I_p \cup O_p)^T - (I_{win} \cup O_{win})^T|$
   $(I_p \cup O_{wp})^T \leftarrow (I_p \cup O_p)^T + \Delta(\varepsilon^T)$
End For

Knowledge transferring for cumulative AI reasoning

TRANSFER$((I_m, O_m), K_{AI})$:
```
EXIST ← FALSE
For all (I_p ∪ O_p)∈K_AI do
   if EXIST = FALSE then
      If (I_p^T = I_m^T) then
         EXIST ← TRUE
         If (O_p^T ≠ O_p^T) then
            INCONSISTENCY ← INCONSISTENCY ∪ ((I_p∪O_p), (I_m∪O_m))
         Enf if
      End if
   End If
End For
If EXIST = FALSE then
   K_AI ← K_AI ∪ (I_m ∪ O_m)
End if
```

Fig. 3. Knowledge translating and transferring principles for knowledge updating.

Next section gives an example of the feasibility of application of such educable learning process by limiting the transfer of K_U to K_{AI} for a cumulative AI system.

5 A Car Driving Use-Case Study

The car driving use-case study concerns a configuration of a crossing between a road with a tram track. It simulates possible users' behaviors when traffic light is red or green and when inter-distance between cars is over or under 2 s. Users' behaviors are limited to the decision of moving or stopping the car. Four rules are then into K_U with regard to these users' behavior integrating three labels (see Fig. 4):

1. The variable with the label "state-light" is for the traffic light state that can be "red" or "green".
2. The variable with the label "state-gap" relates to the inter-distance gap that is "<2 s" or "≥2 s".
3. The variable with the label "state-car" can have the value "stopped" or "moved".

The translation matrix TM contains three triplets. The order of appearance of these triplets in TM is the order of recording of data in K_{AI}. Therefore, the first digit of a triplet is the state of the traffic light, the second one the state of the inter-distance gap and the last one the state of the car.

Initial knowledge bases for educable learning
Rule base K_U related to human reasoning
R1: (state-light = red and state-gap < 2s → state-car= stopped)
R2: (state-light = red and state-gap ≥ 2s → state-car= stopped)
R3: (state-light = green and state-gap < 2s → state-car= stopped)
R4: (state-light = green and state-gap ≥ 2s → state-car= moved)

Knowledge base K_{AI} related to cumulative AI reasoning
{(0 0 0), (0 1 0), (1 0 0), (1 1 1)}

Translation matrix
TM={(state-light, red, green), (state-gap, <2s, ≥2s), (state-car, stopped, moved)}

Fig. 4. Initial bases for educable learning.

Knowledge discovery by users can lead to the production of other rules related to weak signal-based inconsistency occurrence. Such an inconsistency can be the occurrence of a dangerous situation when the traffic light is green and the car decides to move. Indeed, the car will be possibly blocked at the crossing between the road and the tram track if inter-distance between cars becomes under 2 s. A collision can then occur between the car and an arriving tram. Two new labels may be designed. Label "state-position" related to the position of the car may have two values: "on tram track" or "out tram track". Label "state-tram" related to the state of the tram may have same values than the car: "stopped" or "moved". The car may then have an additional value: "collided". This inconsistency discovery makes possible updating of K_U and TM. In order to simplify, this updating maintains initial labels "state-light", "state-gap", "state-car" with two possible values and adds a new one "state-inter" about inter-distance that is under or over 4 s (see Fig. 5). Label "state-inter" appears at the third place on each scenario of K_{AI} for cumulative AI reasoning. Label "state-car" remains with two possible values too.

Educable learning process with an additional rule

Inconsistency discovery based on human reasoning:
(state-car=stopped ∧ state-position=on tram track ∧ state-tram= moved → state-car=collided)
(state-gap ≥ 2s ∧ state-light=green → state-car=not(moved) ⊕ moved)

Updating of rule base K_U related to human reasoning
R1: (state-light = red and state-gap < 2s and state-inter <4s → state-car= stopped)
R2: (state-light = red and state-gap ≥ 2s and state-inter <4s → state-car= stopped)
R3: (state-light = green and state-gap < 2s and state-inter <4s → state-car= stopped)
R4: (state-light = green and state-gap ≥ 2s and state-inter<4s → state-car= stopped)
R5: (state-light = green and state-gap ≥ 2s and state-inter ≥ 4s → state-car= moved)

Updating of knowledge base K_{AI} related to cumulative AI reasoning
{(0 0 0 0), (0 1 0 0), (1 0 0 0), (1 1 0 0), (1 1 1 1)}

New translation matrix
TM={(state-light, red, green), (state-gap, <2s, ≥2s), (state-inter, <4s, ≥4s), (state-car, stopped, moved)}

Fig. 5. Educable learning process

6 Conclusion

This paper has presented challenging issues about human-AI coevolution and shared autonomy between human and AI systems. The educable learning process is proposed as a relevant solution to make possible human and technical knowledge evolution in the course of their activities. It is composed by knowledge discovery and updating by applying characteristics of knowledge representation for human and AI reasoning. Knowledge discovery aims at linking data and identifying relations between them and at exploring new possible relations regarding weak signals-based inconsistencies from users' experience. Knowledge updating consists in translating these relations by respecting knowledge representation constraints of human or AI system and in transferring them adequately. The proposed framework for educable learning process was limited to the discovery and the updating of knowledge by applying natural human reasoning principles and adapting it into merged or cumulative AI system reasoning modalities. A use-case study in car driving domain gave a practical example of its feasibility of application for a cumulative AI system.

Future research will focus on the development of mutual educable learning process between humans and AI systems by implementing features like linkability, explorability, translatability or transferability to optimize knowledge sharing between humans and AI systems. Therefore, the new global feature named educability has to be extended in order to define modalities of human-machine interactions for knowledge exchange, validation, discovery and updating, and make educable learning between humans and AI systems feasible. This has to mutually educate humans and machines to face any inconsistencies by producing online or offline learning and education supports.

Acknowledgement. The present research work has been supported by the Scientific Research Network on Integrated Automation and Human-Machine Systems (GRAISyHM), and by the Regional Council of "Hauts-de-France", project SACRe TIFSID (Technologie Innovante pour un Futur Solidaire, Inclusif, et Durable). The author gratefully acknowledges the support of these institutions.

References

1. Romero, D., Stahre, J., Wuest, T., Noran, O.: Towards an Operator 4.0 typology: a human-centric perspective on the fourth industrial revolution technologies. In: International Conference on Computers & Industrial Engineering (CIE46), At Tianjin, China (2016)
2. Li, S., Abel, M.-H., Negre, E.A.: Collaboration context ontology to enhance human-related collaboration into Industry 4.0. Cogn. Technol. Work **24**, 75–91 (2021)
3. Barcellini, F., et al.: Promises of industry 4.0 under the magnifying glass of interdisciplinarity: revealing operators and managers work and challenging collaborative robot design. Cogn. Technol. Work **25**, 251–271 (2023)
4. Romero, D., Stahre, J.: Towards the resilient operator 5.0: the future of work in smart resilient manufacturing systems. Procedia CIRP **104**, 1089–1094 (2021)
5. Zambiasi, L., Rabelo, R., Zambiasi, S.P., Lizot, R.: Supporting resilient operator 5.0: an augmented softbot approach. In: IFIP Advances in Information and Communication Technology, vol. 664 (2022). https://doi.org/10.1007/978-3-031-16411-8_57
6. Li, X., Nassehi, A., Wang, B., Hu, S.J., Epureanu, B.I.: Human-centric manufacturing for human-system coevolution in Industry 5.0. CIRP Ann. **72**(1), 393–396 (2023)
7. Vanderhaegen, F.: A non-probabilistic prospective and retrospective human reliability analysis method—application to railway system. Reliab. Eng. Saf. Syst. **71**(1), 1–13 (2001)
8. Vanderhaegen, F.: Human-error-based design of barriers and analysis of their uses. Cogn. Technol. Work **12**, 133–142 (2010)
9. Hickling, E.M., Bowie, J.E.: Applicability of human reliability assessment methods to human–computer interfaces. Cogn. Technol. Work **15**, 19–27 (2013)
10. Rangra, S., Sallak, M., Schön, W., Vanderhaegen, F.: A graphical model based on performance shaping factors for assessing human reliability. IEEE Trans. Reliab. **66**(4), 1120–1143 (2017)
11. de Winter, J.C.F., Eisma, Y.B., Cabrall, C.D.D., Hancock, P.A., Stanton, N.A.: Situation awareness based on eye movements in relation to the task environment. Cogn. Technol. Work **21**(1), 99–111 (2018). https://doi.org/10.1007/s10111-018-0527-6
12. Vanderhaegen, F., Wolff, M., Mollard, R.: Non-conscious errors in the control of dynamic events synchronized with heartbeats: a new challenge for human reliability study. Saf. Sci. **129**, 104814 (2020)
13. Fürstenau, N., Radüntz, T.: Power law model for subjective mental workload and validation through air traffic control human-in-the-loop simulation. Cogn. Technol. Work **24**, 291–315 (2022)
14. Vanderhaegen, F., Wolff, M., Mollard, R.: A heartbeat-based study of attention in the detection of digital alarms from focused and distributed supervisory control systems. Cogn. Technol. Work **25**, 119–134 (2023)

15. Navarro, J., Jomard, E., Saleur, É., Abrard, W., Cegarra, J.: Influence of automation mode use on selection rates and subjective assessment over time. Cogn. Technol. Work **24**, 609–624 (2022)
16. Schwarz, C., Gaspar, J., Brown, T.: The effect of reliability on drivers' trust and behavior in conditional automation. Cogn. Technol. Work **21**, 41–54 (2019)
17. Mougi, A.: The roles of amensalistic and commensalistic interactions in large ecological network stability. Sci. Rep. **6**, 29929 (2016). https://doi.org/10.1038/srep29929
18. Xu, P.: Dynamics of microbial competition, commensalism, and cooperation and its implications for coculture and microbiome engineering. Biotechnol. Bioeng (2020). https://doi.org/10.1002/bit.27562
19. Helmenstine, A.: Symbiosis Definition and Examples. Science Notes Posts, February 11, 2023 (2023). https://sciencenotes.org/symbiosis-definition-and-examples/
20. Cheam, D., Nishiguchi, M.K.: Symbiosis: a review of different forms of interactions among organisms. Encyclopedia Biodiversity (Third Edition) **6**, 242–252 (2024)
21. Zieba, S., Polet, P., Vanderhaegen, F., Debernard, S.: Principles of adjustable autonomy: a framework for resilient human machine cooperation. Cogn. Technol. Work **12**(3), 193–203 (2010)
22. Ouedraogo, K.-A., Enjalbert, S., Vanderhaegen, F.: How to learn from the resilience of Human-Machine Systems? Eng. Appl. Artif. Intell. **26**(1), 24–34 (2013)
23. Vanderhaegen, F.: Towards increased systems resilience: new challenges based on dissonance control for human reliability in Cyber-Physical & Human Systems. Annu. Rev. Control. **44**, 316–322 (2017)
24. Enjalbert, S., Vanderhaegen, F.: A hybrid reinforced learning system to estimate resilience indicators. Eng. Appl. Artif. Intell. **64**, 295–301 (2017)
25. Vanderhaegen, F.: Cooperative system organisation and task allocation: illustration of task allocation in air traffic control". Le Travail Humain **62**, 197–222 (1999)
26. Vanderhaegen, F.: Toward a model of unreliability to study error prevention supports. Interact. Comput. **11**, 575–595 (1999)
27. Vanderhaegen, F.: Human Error Analysis and Control. Hermès Science Publications, Paris, France (2003)
28. Carsten, O., Martens, M.H.: How can humans understand their automated cars? HMI principles, problems and solutions. Cogn. Technol. Work **21**, 3–20 (2019)
29. Tausch, A., Kluge, A.: The best task allocation process is to decide on one's own: effects of the allocation agent in human–robot interaction on perceived work characteristics and satisfaction. Cogn. Technol. Work **24**, 39–55 (2022)
30. Endsley, M.R.: Toward a theory of situation awareness in dynamic systems. Hum. Factors **37**(1), 32–64 (1995)
31. Carsten, O., Vanderhaegen, F.: Situation awareness: valid or fallacious? Cogn. Technol. Work **17**, 157–158 (2015)
32. Salmon, P.M., Walker, G.H., Stanton, N.A.: Broken components versus broken systems: why it is systems not people that lose situation awareness. Cogn. Technol. Work **17**(2), 179–183 (2015)
33. Dekker, S.W.A.: The danger of losing situation awareness. Cogn. Technol. Work **17**(2), 159–161 (2017)
34. Vanderhaegen, F., Wolff, M., Mollard, R.: Repeatable effects of synchronizing perceptual tasks with heartbeat on perception-driven situation awareness. Cogn. Syst. Res. **81**, 80–92 (2023)
35. Weick, K.E.: Sensemaking in Organizations. Sage, Thousand Oaks (1995)
36. Jensen, E.: Sensemaking in military planning: a methodological study of command teams. Cogn. Technol. Work **11**, 103–118 (2009)

37. Albolino, S., Cook, R., O'Connor, M.: Sensemaking, safety, and cooperative work in the intensive care unit. Cogn. Technol. Work **9**, 131–137 (2007)

38. Roscher, R., Bohn, B., Duarte, M.F., Garcke, J.: Explainable machine learning for scientific insights and discoveries. IEEE Access **8**, 42200–42216 (2020)

39. Paleja, R., Ghuy, M., Arachchige, N.R., Jensen, R., Gombolay, M.: The utility of explainable AI in ad hoc human-machine teaming. In: 35th Conference on Neural Information Processing Systems (NeurIPS 2021), 6–14 December (2021)

40. Srivastava, D., Lilly, J.M., Feigh, K.M.: Exploring the role of judgement and shared situation awareness when working with AI recommender systems. Cogn. Technol. Work (2024). https://doi.org/10.1007/s10111-024-00771-9

41. Hauptman, A.I., Schelble, B.G., Duan, W., et al.: Understanding the influence of AI autonomy on AI explainability levels in human-AI teams using a mixed methods approach. Cogn. Technol. Work (2024). https://doi.org/10.1007/s10111-024-00765-7

42. Harrison, R., Flood, D., Duce, D.: Usability of mobile applications: literature review and rationale for a new usability model. J. Interact. Sci. (2013). https://doi.org/10.1186/2194-082 7-1-1

43. Joo, S., Lin, S., Lu, K.: A usability evaluation model for academic library websites: efficiency, effectiveness and learnability. J. Libr. Inf. Stud. **9**(2), 11–26 (2011)

44. Sheridan, T.B.: Telerobotics, Automation, and Human Supervisory Control. MIT Press, Cambridge (1992)

45. Inagaki, T.: Design of human–machine interactions in light of domain-dependence of human-centered automation. Cogn. Technol. Work **8**, 161–167 (2006)

46. Beer, J., Fisk, A., Rogers, W.: Toward a framework for levels of robot autonomy in human-robot interaction. J. Hum.-Robot Interact. **3** (2014). https://doi.org/10.5898/JHRI.3.2.Beer

47. Inagaki, T., Sheridan, T.B.: A critique of the SAE conditional driving automation definition, and analyses of options for improvement. Cogn. Technol. Work **21**, 569–578 (2018)

48. Merat, N., et al.: The "Out-of-the-Loop" concept in automated driving: proposed definition, measures and implications. Cogn. Technol. Work **21**, 87–98 (2019)

49. Brandenburger, N., Naumann, A., Jipp, M.: Task-induced fatigue when implementing high grades of railway automation. Cogn. Technol. Work **23**, 273–283 (2021)

50. Vanderhaegen, F.: Pedagogical learning supports based on human–systems inclusion applied to rail flow control. Cogn. Technol. Work **23**, 193–202 (2021)

51. Vanderhaegen, F., Nelson, J., Wolff, M., Mollard, R.: From human-systems integration to human-systems inclusion for use-centred inclusive manufacturing control systems. IFAC-PapersOnLine **54**(1), 249–254 (2021)

52. Vanderhaegen, F., Jimenez, V.: Opportunities and threats of interactions between humans and cyber–physical systems – integration and inclusion approaches for CPHS. In: Annaswamy, A.M., Khargonekar, P.P., Lamnabhi-Lagarrigue, F., Spurgeon, S.K. (eds.) Cyber–Physical–Human Systems: Fundamentals and Applications, pp. 71–90. Wiley (2023)

53. Bonet-Jover, A., Sepúlveda-Torres, R., Saquete, E., Martínez-Barco, P.: A semi-automatic annotation methodology that combines summarization and human-in-the-loop to create dis-information detection resources. Knowl.-Based Syst. **275** (2023). https://doi.org/10.1016/j.knosys.2023.110723

54. Delussu, R., Putzu, L., Fumera, G.: Human-in-the-loop cross-domain person re-identification. Expert Syst. Appl. **226** (2023). https://doi.org/10.1016/j.eswa.2023.120216

55. Mosqueira-Rey, E., Hernández-Pereira, E., Alonso-Ríos, D., Bobes-Bascarán, J., Fernández-Leal, A.: Human-in-the-loop machine learning: a state of the art. Artif. Intell. Rev. **56**, 3005–3054 (2023)

56. Fu, L.M.: Knowledge discovery based on neural networks. Commun. ACM **42**(11), 47–50 (1999)

57. Shu, X., Ye, Y.: Knowledge discovery: methods from data mining and machine learning. Soc. Sci. Res. **110** (2023). https://doi.org/10.1016/j.ssresearch.2022.102817

58. Ahlqvist, T., Uotila, T.: Contextualising weak signals: towards a relational theory of futures knowledge. Futures **119** (2020). https://doi.org/10.1016/j.futures.2020.102543

59. Vanderhaegen, F.: Weak signal-oriented investigation of ethical dissonance applied to unsuccessful mobility experiences linked to human–machine interactions. Sci. Eng. Ethics **27**, 2 (2021). https://doi.org/10.1007/s11948-021-00284-y

60. Vanderhaegen, F.: Dissonance engineering: a new challenge to analyse risky knowledge when using a system. Int. J. Comput. Commun. Control **9**(6), 776–785 (2014)

61. Vanderhaegen, F.: A rule-based support system for dissonance discovery and control applied to car driving. Expert Syst. Appl. **65**, 361–371 (2016)

62. Vanderhaegen, F.: Heuristic-based method for conflict discovery of shared control between humans and autonomous systems - a driving automation case study. Robot. Auton. Syst. **146** (2021). https://doi.org/10.1016/j.robot.2021.103867

63. Schieben, A., Wilbrink, M., Kettwich, C., Madigan, R., Louw, T., Merat, N.: Designing the interaction of automated vehicles with other traffic participants: design considerations based on human needs and expectations. Cogn. Technol. Work **21**, 69–85 (2019)

64. Vanderhaegen, F., Zieba, S., Polet, P.: A reinforced iterative formalism to learn from human errors and uncertainty. Eng. Appl. Artif. Intell. **22**(4–5), 654–659 (2009)

65. Vanderhaegen, F., Zieba, S., Polet, P., Enjalbert, S.: A Benefit/Cost/Deficit (BCD) model for learning from human errors. Reliab. Eng. Syst. Saf. **96**(7), 757–766 (2011)

66. Vanderhaegen, F., Zieba, S.: Reinforced learning systems based on merged and cumulative knowledge to predict human actions. Inf. Sci. **276**, 146–159 (2014)

67. Sternberg, R.J.: Toward a unified theory of human reasoning. Intelligence **10**(4), 281–314 (1986)

68. Evans, J.S.B.T.: Logic and human reasoning: an assessment of the deduction paradigm. Psychol. Bull. **128**(6), 978–996 (2002)

69. Jouglet, D., Piechowiak, S., Vanderhaegen, F.: A shared workspace to support man-machine reasoning: application to cooperative distant diagnosis. Cogn. Technol. Work **5**, 127–139 (2003)

70. Vanderhaegen, F., Jouglet, D., Piechowiak, S.: Human-reliability analysis of cooperative redundancy to support diagnosis. IEEE Trans. Reliab. **53**, 458–464 (2004)

71. Vanderhaegen, F., Caulier, P.: A multi-viewpoint system to support abductive reasoning. Inf. Sci. **181**, 5349–5363 (2011)

72. Aguirre, F., Sallak, M., Vanderhaegen, F., Berdjag, D.: An evidential network approach to support uncertain multiviewpoint abductive reasoning. Inf. Sci. **253**, 110–125 (2013)

73. Van Hoeck, N., Watson, P.D., Barbey, A.K.: Cognitive neuroscience of human counterfactual reasoning. Front. Hum. Neurosci. (2015). https://doi.org/10.3389/fnhum.2015.00420

74. Stephens, R.G., Dunn, J.C., Hayes, B.K.: Are there two processes in reasoning? The dimensionality of inductive and deductive inferences. Psychol. Rev. **125**(2), 218–244 (2018)

Artificial Intelligence for Human Resource Management: A Review, Perspective, and Case Study

Dac Hieu Nguyen[1,2], Kim Duc Tran[1,3](✉), Nguyen Anh Luong[1,4],
Minh Anh Luong[1,5], Sébastien Thomassey[1,3], and Kim Phuc Tran[1,3]

[1] International Chair in DS and XAI, International Research Institute for Artificial Intelligence and Data Science, Dong A University, Danang, Vietnam
`ductk@donga.edu.vn`
[2] Institute of Artificial Intelligence and Data Science, Thuyloi University, Hanoi, Vietnam
[3] Univ. of Lille, ENSAIT, ULR 2461 - GEMTEX - Génie et Matériaux Textiles, 59000, Lille, France
[4] Human Resources Administration Office, Dong A University, Danang, Vietnam
[5] Faculty of Business Management, Dong A University, Danang, Vietnam

Abstract. Integrating artificial intelligence (AI) into Human Resource Management (HRM) has become a critical advancement in maximizing HR activities, from recruiting and performance assessment to employee retention and organizational growth. Emphasizing the transforming power of artificial intelligence in strengthening decision-making processes, increasing efficiency, and thus advancing justice in HR practices, this study investigates the present situation of AI applications in HRM. Our research also covers the ethical issues related to artificial intelligence, like prejudice, openness, and respect for employee dignity. Additionally, we present a case study on implementing an AI-based Decision Support System (DSS) to predict employee attrition, showcasing its potential to significantly enhance retention strategies and overall organizational stability. This paper offers a complete view of the possibilities and difficulties of artificial intelligence in HRM by analyzing the most recent studies and pragmatic implementations. The results highlight the importance of responsible artificial intelligence use in line with ethical norms and human values. This guarantees that artificial intelligence will be a tool for enhancing rather than replacing human HRM decision-making.

Keywords: Human Resource Management · HR decision-making · AI-based Decision Support Systems · data-driven analysis · Ethical AI

1 Introduction

Within numerous domains, such as Human Resource Management (HRM), artificial intelligence (AI) has a significant and influential impact. The need for

S. Thomassey et al. (Eds.): RAIDS 2024, LNICST 673, pp. 124–138, 2026.
https://doi.org/10.1007/978-3-032-14055-5_9

improved decision-making quality, increased efficiency, and managing the growing complexity of human capital in the current fast-paced corporate environments has driven the integration of AI into HRM approaches. Traditional human resource management functions such as hiring, assessing individuals, managing performance, and developing organizations are being revolutionized by the potential of artificial intelligence to automate repetitive work, offer data-driven analysis, and enhance decision accuracy [26]. The fast development of AI technologies has made it possible to build sophisticated tools and systems to manage large amounts of data to assist HR decision-making effectively, particularly in machine learning, natural language processing, and data analytics. These technologies improve the efficiency and accuracy of HR operations and provide fair and tailored outcomes by the general goals of corporate performance and employee satisfaction [28]. The duties of human resource managers are moving from administrative chores to strategic decision-making roles as businesses adopt AI-driven HRM solutions more and more. In this sense, they enable AI technologies in the workplace [8]. Still, the integration of AI into HRM is not without challenges. Ensuring the correct use of AI in HRM calls for resolving ethical issues, including bias in AI algorithms, transparency in decision-making, and safeguarding of employee dignity, which are all related. Furthermore, significant challenges to the broad use of AI technology in this field are the fear of automation and the prospect of job displacement [18]. Therefore, knowing the possible benefits and challenges connected with AI in HRM is important to create effective plans that maximize AI possibilities and reduce risks. We aim to thoroughly assess AI applications in HRM from the current state. This paper will include the most current developments in research, useful applications, and the moral implications of using AI in this field. The confluence of AI and HRM is investigated in this paper to provide an insightful analysis of how AI may be used to enhance HRM procedures. It underlines the importance of using AI to support respect for human values, justice, and openness [24].

Our remaining research is arranged in the following sequence: Emphasizing its use in enhancing decision-making by employing AI and large data, Sect. 2 explores the Decision Support System (DSS) in HRM. Emphasizing neural networks and text mining methods, Sect. 3 investigates applying AI in HRM. These technologies improve hiring procedures and raise staff output. Section 4 addresses the moral conundrums of deploying AI in HRM by emphasizing prejudice, openness, and maintaining employee dignity. Emphasizing reaching a harmonic balance between effectiveness and moral values, Sect. 5 investigates the difficulties and possible future advancements of integrating AI into HRM. Section 6 presents a case study on an AI-driven Decision Support System (DSS) in HRM, highlighting its pragmatic applications and consequences. Section 7 of the chapter summarizes significant findings and recommends areas for more research on the subject.

2 Decision Support System in Human Resource Management

Human resource management (HRM) decisions often suffer from biases and inconsistencies due to reliance on human judgment. Decision Support Systems (DSS) address these challenges by providing data-driven insights, enabling more objective and consistent decision-making. By integrating advanced technologies like artificial intelligence (AI), DSS enhances the effectiveness of HRM, ensuring that decisions align with organizational goals and contribute to overall success [1]. This section provides an overview of DSS in HRM and the current research progress in this field.

In Industry 4.0, the emergence of big data has significantly transformed HRM, leading to the development of advanced DSS. Mert Bal et al. [3] introduced a novel decision support system for improving the recruitment process in HRM by leveraging based on formal contexts and concept lattices. Formal Concept Analysis (FCA) is a theoretical framework for mathematical data analysis. This approach creates a concept lattice using candidate qualifications and job specifications to extract exact and approximate association rules through association rule mining, supporting HR managers and practitioners in handling huge applicant numbers for vacancies. This matching will expedite the process and quickly locate the most qualified candidates before interviews. Samuel et al. [21] leverages fuzzy logic, a powerful soft computing tool, to model decision-making processes that account for human evaluations' inherent uncertainties and subjectivities. The system provides a more objective and consistent appraisal process by incorporating key performance attributes such as academic qualifications, research output, and teaching experience. This approach used the Mamdani fuzzy inference method and Center of Gravity. Defuzzification ensures that the system can handle complex decision variables and deliver precise performance evaluations, showing that the proposed approach made 78% correct evaluation predictions. To solve big data challenges, Chungling Cai and Chuanyi Chen proposed a Platform for Decision Support in Human Resource Information Management, using cloud computing to effectively handle large and varied data sources within human resource systems. [8]. The architecture consists of seven layers: the Data Acquisition Layer, which collects different types of data; the Network Service Support Layer, which enables adaptive data transmission; the Cloud Computing Support Layer, which facilitates cloud-based processing; the Data Standardization Transformation Layer, which unifies diverse data; the System Application Layer, which provides HR file management services; the System Service Layer, which integrates resources; and the Decision Support Layer, which conducts data mining and analysis to assist in decision-making. The platform's design facilitates efficient data mining, processing, and interpretation by utilizing computing, storage, and data resource pools. It also includes an analysis platform that categorizes, processes, and interprets data to generate strategic and responsive reports, thereby improving the accuracy and effectiveness of HR decisions.

Due to traditional DSS uses static models and predetermined rules, they often struggle in complex and uncertain business environments. Intelligent Decision Support Systems (IDSS) come into play [14]. Integrating modeling tools with human experience enables these systems to effectively manage the growing intricacy and quantity of HR data, hence assuring more precise and reliable decision-making processes. Based on the findings of Ireneusz Czarnowskia et al. [10], machine learning algorithms such as CART, C 4.5, and kNN have been effectively implemented to monitor and evaluate HR activities, predicting decisions that could optimize HR processes and reduce the impact of undesirable behaviors within an organization. In computational experiments, algorithms such as C4.5 with pruning have achieved classification accuracies of up to 84.32%, indicating their potential to provide reliable and precise decision-making support in HRM.

With the increased availability and accessibility of HR data, HR practitioners may better use HR metrics to measure HR's efficiency, effectiveness, impact, and service. HR metrics are categorized into four levels [11]: efficiency metrics, which assess productivity and cost; human capital metrics, which measure the value of employees; effectiveness metrics, which evaluate the success of HR programs; and strategic impact metrics, which demonstrate HR's influence on broader business outcomes. Despite the development of numerous HR metrics, there is a lack of clear guidance on how and where to apply these metrics. To address this issue, James H. Dulebohn and Richard D. Johnson [11] proposed a framework that guides the application of HR metrics and DSS tools across different HR activities, emphasizing the importance of aligning these tools with the specific decision-making needs at various levels of organizational functioning and also outlines propositions for future research to refine further the use of HR metrics, DSS, and BI capabilities, aiming to support HR's strategic role in organizations better.

3   Artificial Intelligence in Human Resources Management Practices

Within the context of Industry 4.0, incorporating Artificial Intelligence (AI) into all aspects of business operations is progressively advancing. Hence, the utilization of AI in Human Resource Management (HRM) is emerging as a strategic field with significant potential for innovation and enhanced efficiency. Some examples of applications include artificial neural networks for predicting employee turnover, knowledge-based search engines for retrieving candidate information, genetic algorithms for scheduling staff, text mining for analyzing sentiment, information extraction for obtaining resume data, and interactive voice response systems for self-service by employees [26]. This section will summarize the research and uses of AI in HRM, focusing on the latest advancements and implementations in this field.

Nishad Nawaz et al. [20] conducted a recent study polling 274 IT workers in Chennai. Among the respondents, 55.5% were female, and 51.5% were between the ages of 26 and 30. Most (39.1%) possessed professional degrees, whereas 81.4% were married. Most respondents reported an income between Rs 25,001 and Rs 50,000, while 37.6% had 2–5 years of industry experience. The study employed Structural Equation Modeling (SEM) [15] to investigate the correlations among several AI performance metrics, such as accuracy, automation, computing power and capacity, real-time experience, and personalization, and their subsequent effect on optimizing time efficiency and cost reduction. Therefore, no substantial evidence was discovered to affirm the influence of computing power, capacity, and personalization on improving time efficiency and reducing expenses. Likewise, Murugesan and colleagues [19] surveyed 271 HR experts from different industries, achieving a response rate of 75%. This study explored AI's influence on human resources operations, focusing on employee productivity, health and safety management, payroll processing, and the provision of real-time feedback. This study, which also used SEM, established that AI greatly improves the agility of human resources (HR), especially in the analysis and design of organizational networks. The coefficient value for health and safety improvements was determined to be 0.660, while employee productivity assessment contributed 0.422 to the digitization of HR. The adoption of AI in HRM was investigated by Sithambaram and Tajudeen in 2022 [23]. The study included 12 firms in Malaysia representing various sectors like healthcare, IT, and finance, with employee numbers ranging from 18 to 835,000. According to the study's qualitative analysis of interviews with HR experts, AI is largely employed in recruiting, with platforms such as Taleo, Workday, and Spire AI automatically eliminating 75% of unqualified CVs, significantly reducing hiring time and enhancing recruiter productivity. Additionally, this research reveals that 10 out of 12 companies experienced organizational impacts such as employee empowerment, improved morale, and the ability to focus on value-added tasks. Furthermore, AI's usage in compliance processes resulted in a 16% rise in new-hire diversity across five organizations. Giving concrete evidence of AI's transformative impact on HRM in Malaysia, highlighting its potential to enhance efficiency, accuracy, and strategic decision-making in the HR sector. To optimize operations and reduce the effort of HRM, Song and Wu [24] introduce a backpropagation neural network (BPNN)-based HRM system model that aims to optimize HR processes by increasing efficiency, decreasing workload, and enhancing predictive accuracy. The model leverages DL algorithms to facilitate faster convergence and more accurate predictions, such as salary forecasting, compared to traditional HRM models. Specifically, the BPNN-based HRM system model achieved a prediction accuracy of 88.72%, which is 2.76% higher than traditional models, and convergence in approximately 60 epochs and demonstrates that it significantly outperforms other models in terms of both speed and accuracy, offering a robust solution for the intelligent development of HRM systems in modern enterprises.

The advent of generative AI has emerged as a transformative trend in AI, with ChatGPT at the forefront, significantly impacting HRM practices [6]. ChatGPT was reshaping HRM by automating and enhancing various functions such as recruitment, onboarding, training, and performance management. ChatGPT can screen resumes, conduct initial interviews, and deliver personalized feedback, improving efficiency and engagement. Notably, with ChatGPT, employees no longer have to wait for Human Resources representatives to respond to their questions, as they can receive immediate, tailored answers, significantly improving the employee experience [17]. Beyond these operational benefits, ChatGPT facilitates continuous learning by recommending customized training programs tailored to individual employee needs, thereby supporting professional development.

Looking ahead to future research, Chowdhury et al. [9] present a comprehensive AI capability framework for integrating AI into HRM, drawing on the Resource-Based View (RBV) and Knowledge-Based View (KBV) theories and emphasizing the need for both technical (e.g., data management, AI algorithms) and non-technical resources (e.g., leadership, culture, knowledge management) to maximize AI's potential in HRM. The framework also identifies key areas for future research, including exploring the impact of organizational resources on business outcomes, prioritizing resources for AI adoption, enhancing AI transparency, fostering AI-employee collaboration in HRM, focusing on bias and fairness, explainability and transparency, and preserving HRM to achieve sustained business performance.

4   The Ethical Use of Artificial Intelligence in Human Resource Management

The rapid rise of Artificial Intelligence (AI) in workplaces offers significant potential for enhancing business profitability. However, implementing AI in Human Resource Management (HRM) without careful consideration could lead to unintended consequences, posing serious ethical questions about data privacy, equity, and the suitability of automated decisions that directly impact individuals' careers. Despite the importance of these concerns, research in HRM and technology often focuses on AI's capabilities rather than its ethical use or how to involve human workers effectively. This section delves into the critical ethical challenges of integrating AI in HRM, focusing on bias and fairness, explainability and transparency, and preserving workers' dignity.

4.1   Bias and Fairness

The growing use of AI in HRM raises critical concerns regarding bias and fairness, especially in recruitment. Recruitment algorithms often perpetuate historical biases, such as underrepresenting women in technical roles due to biased training data [4]. Algorithmic focus bias, where inappropriate data influences decisions, further exacerbates this issue, as seen in cases where commuting times were unfairly used to disadvantage applicants from minority neighborhoods. In the study by Köchling et al. [18], despite Asians having higher job interview scores in both the training and test datasets, the algorithms significantly underestimated their likelihood of being invited for a job interview. This underestimation, driven by underrepresentation in the data set, led to unpredictable classifications, ultimately reducing their chances of being selected for interviews despite having equal qualifications. Emphasizing the importance of carefully evaluating and managing AI systems to achieve fair and equitable results in HRM practices.

4.2 Explainability and Transparency

In HRM, especially in data-driven decision-making, limited transparency and explainability in artificial intelligence systems pose major difficulties. As AI technologies become more integrated into HRM practices, algorithms' proprietary or complex often make it difficult for HR managers to understand how AI-generated insights are derived, which can undermine trust in these tools and raise ethical concerns. For example, during employee recruitment, if an AI algorithm results in an unfavorable outcome for an applicant, neither the applicant nor the HR managers may clearly understand the reasoning behind the decision [9]. Lack of transparency can lead to accountability concerns, reduced confidence in AI-driven decisions, and decreased employee confidence. To tackle these concerns, the Transparency by Design (TbD) framework has been proposed [29], embedding accountability, system design, and information analysis into AI development to ensure transparent and fair practices. Additionally, Explainable AI (XAI) techniques, such as interpretable models and post-hoc analysis tools such as LIME and SHAP [12], are essential for making AI decisions more understandable and trustworthy. Integrating these approaches into HRM makes it possible to balance accuracy with transparency, fostering more ethical and effective management practices while ensuring that AI tools are trusted and accepted within HRM.

4.3 Workers' Dignity

Since AI is unable to comprehend or care about human feelings, it is difficult to treat employees with the respect they deserve, leading to scenarios where workers are treated as mere objects or means to an end, thus undermining their dignity [28]. This situation is exacerbated when AI makes decisions that significantly impact workers' lives, such as hiring, promotions, or performance evaluations, without the necessary human oversight to ensure fairness and empathy. Bankins et al. [5] surveyed to investigate how employees perceive fairness, trust, and dehumanization in HRM decisions made by AI versus humans across various HR functions. It involved 446 diverse North American participants who evaluated scenarios where decisions were either positive or negative. Findings revealed that human decision-makers were typically preferred, particularly regarding perceived justice and trust. In contrast, AI decisions were seen as more data-driven but often less appropriate and more dehumanizing, especially when outcomes were negative. To protect workers' dignity, it is crucial to implement AI systems that are properly controlled and integrated with human judgment, ensuring that they enhance rather than diminish the humanity of the workplace.

5 Difficulties, Challenges, and Perspectives for Artificial Intelligence on Human Resources Management

Integrating Artificial Intelligence (AI) into Human Resource Management (HRM) presents promising opportunities and significant challenges that must be carefully explored to realize its full potential. As organizations increasingly adopt AI technologies to enhance HRM practices, it is crucial to understand the difficulties and challenges accompanying this technological evolution and explore perspectives investigating potential future developments in this field.

One of the foremost challenges is the fear of automation, which has historically hindered the adoption of new technologies in workplaces [16,30]. This fear is particularly acute in HRM, where there is concern that AI could displace human roles, especially in tasks traditionally reliant on human judgment, such as talent acquisition and employee evaluations. As a result, organizations must prioritize transparency and communication to help employees understand how AI will impact their roles and to mitigate anxiety about job displacement. Furthermore, offering employees the opportunity to reskill and upskill can assist them in adjusting to new technologies, ensuring that AI is intended to enhance their work rather than replace it. Organizations can mitigate concerns and establish a more positive environment for AI adoption by emphasizing collaboration between humans and machines. Human-centered HRM-AI [7] addresses these concerns by ensuring that AI technologies are implemented as tools to augment, rather than replace, human roles. Emphasizing the importance of maintaining human oversight in decision-making processes, thereby preserving the critical elements of human judgment and empathy in HR tasks like recruitment and employee evaluations. In the future, this field of study is expected to progress with an increasing focus on developing AI systems that are not only technologically sophisticated but also compatible with ethical standards and human values.

Despite its potential to improve employment efficiency, AI also presents substantial concerns such as reinforcing discrimination, invading privacy, and reducing human involvement in decision-making, as discussed in Sect. 4. However, Hunkenschroer and Kriebitz [13] argue that these risks are not intrinsic to AI but depend on its implementation. They propose that with appropriate design, regular auditing, and deliberate integration, AI can be used ethically in recruitment, complementing human judgment rather than replacing it. Ultimately, AI in recruitment is not inherently unethical but requires organizations to ensure that these tools respect human rights, promote fairness, and maintain transparency. Additionally, addressing privacy concerns is crucial, as AI's usage of employee data, especially from social media and digital traces, can lead to perceived invasions of privacy [27]. To mitigate this, enterprises must adopt privacy-preserving technologies and comply with regulations such as the European Union's (EU) General Data Protection Regulation (GDPR), ensuring that AI tools respect human rights, promote fairness, and maintain transparency. Despite the importance of privacy, there remains a lack of comprehensive empirical research on this topic, indicating a critical gap that future studies must address to ensure responsible AI deployment in HRM [7]. Ultimately, AI in recruitment is not inherently unethical but requires careful management to balance efficiency with ethical considerations.

The efficient operation of AI systems is highly dependent on massive amounts of high-quality data. However, HR data is often fragmented, outdated, or incomplete in many organizations. This poses a significant challenge to the successful implementation of AI in HRM. Without accurate and comprehensive data [9,27], AI algorithms may produce unreliable results, leading to poor decision-making. Ensuring data integrity and establishing robust data governance frameworks are essential steps that organizations must take to support AI-driven HR initiatives

Looking ahead, the successful integration of AI into HRM will require a concerted effort to address these challenges. Organizations need to develop strategies for managing the ethical implications of AI, ensuring that AI systems are transparent, fair, and aligned with broader societal values. Additionally, HR professionals will need to acquire new skills in data analysis and AI technologies to effectively leverage these tools in their work. Overcoming opposition and promoting innovation will depend on creating a supportive workplace culture that welcomes AI as a complementing tool instead of a threat.

6   Case Study: AI-Based Decision Support System in Human Resource Management

In this section, we conduct a case study that explores the application of Artificial Intelligence (AI) in predicting employee attrition, a critical challenge organizations face. By leveraging AI to predict employee attrition, we hope organizations can enhance retention strategies, preserve staff stability, and achieve better organizational outcomes.

Data Preparation

The Human Resource Management (HRM) dataset used in this investigation was obtained from Kaggle [22] and consists of 8 features, as described in Table 1, spread over 1,000 observations. The dataset analyzes employee attrition and can help the organization implement strategies to improve employee retention, optimize human resource practices, and maintain a stable workforce.

Methodology

Originally developed for natural language processing tasks, the Transformer model demonstrated exceptional performance in modeling sequential data due to its ability to capture long-range dependencies. In this study, we adapt the Transformer model to process structured tabular data, as illustrated in Fig. 1. The Transformer model is a neural network architecture that uses self-attention

Table 1. Dataset Feature Descriptions

Feature	Description
EmployeeID	Unique employee identifier
TotalMonthsOfExp	Total experience in months
TotalOrgsWorked	Number of organizations worked in
MonthsInOrg	Months in organization
LastPayIncrementBand	Last pay increment
AverageFeedback	Average performance feedback score
LastPromotionYears	Years since last promotion
Attrition	Indicator of employee turnover (1 = Yes, 0 = No)

mechanisms to process input sequences in parallel, consisting of an encoder, which converts the input into a dense representation, and a decoder, which generates the output sequence.

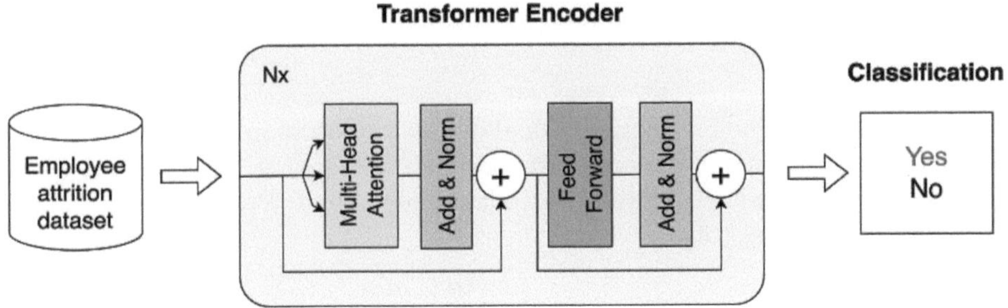

Fig. 1. The illustration of a Transformer for employee attrition classification

Experimental Result

We divided the dataset into training, validation, and test sets with a 50:20:30 ratio. To achieve optimal performance, careful tuning of the hyperparameters was essential. The final batch size selected for our architecture was 32, and we employed the Adam optimizer. The multi-head attention mechanism was configured with 2 heads for effective feature learning. The experiment was implemented with PyTorch version 2.3.1 and Python version 3.11.9 on a computer using a chipset Intel i5-13500 processor with NVIDIA GeForce RTX 3060 GPU acceleration, 512 GB SSD, and 32 GB of RAM.

The evaluation of the model is conducted using four essential metrics: Accuracy, Recall, Precision, and F1 Score. Accuracy is a quantitative measure that reflects the frequency with which a model correctly forecasts a certain result. Precision is the ratio of correctly predicted positive outcomes to all positive predictions. Recall quantifies the proportion of accurate positive predictions from all

Table 2. The experimental results

Model	Accuracy	Recall	Precision	F1 Score
Transformer	0.9885	0.9620	0.9870	0.9743

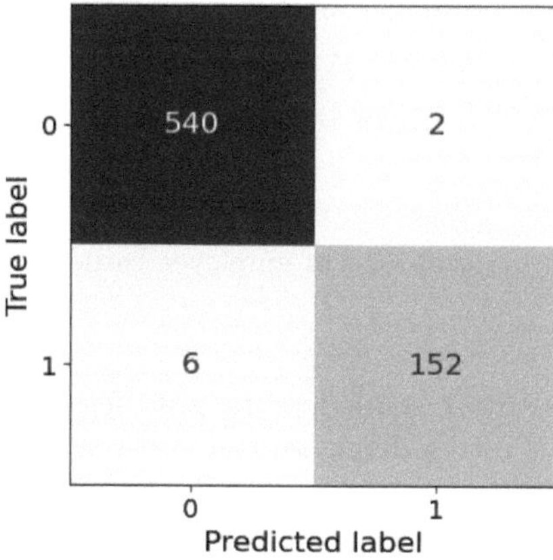

Fig. 2. Confusion Matrix for Transformer model

positive instances. The F1 Score is calculated as the harmonic mean of accuracy and recall. It efficiently balances high precision and strong recall by penalizing extremely negative numerical values in either component. Presented here are the mathematical representations:

$$Accuracy = \frac{TP + TN}{TP + TN + FP + FN} \tag{1}$$

$$Recall = \frac{TP}{TP + FN} \tag{2}$$

$$Precision = \frac{TP}{TP + FP} \tag{3}$$

$$F1\ Score = \frac{2 \times Precision \times Recall}{Precision + Recall} \tag{4}$$

The abbreviations TP, TN, FP, and FN represent the number of true positives, true negatives, false positives, and false negatives, respectively.

The experimental results demonstrate the exceptional performance of the Transformer model in predicting employee attrition, shown in Table 2. With an impressive accuracy of 98.85%, successfully identifying the correct class in nearly all cases. F1 Score of 97.43%, indicating overall model overall effectiveness and robustness. We also made the confusion matrix in Fig. 2 visually support these metrics to analyze the difference between the predicted and the actual value, indicating a very low false negative rate, reflecting the model's strong ability to identify non-attrition cases correctly.

The corresponding training and validation graphs provide additional insights into the model's learning process, as Fig. 3 illustrates. The training and validation loss curves consistently decrease over the epochs, converging to near-zero values by the 20th epoch. The training and validation accuracy graphs steadily increase, stabilizing around 98–99% accuracy, confirming the model's robust generalization capabilities on unseen data.

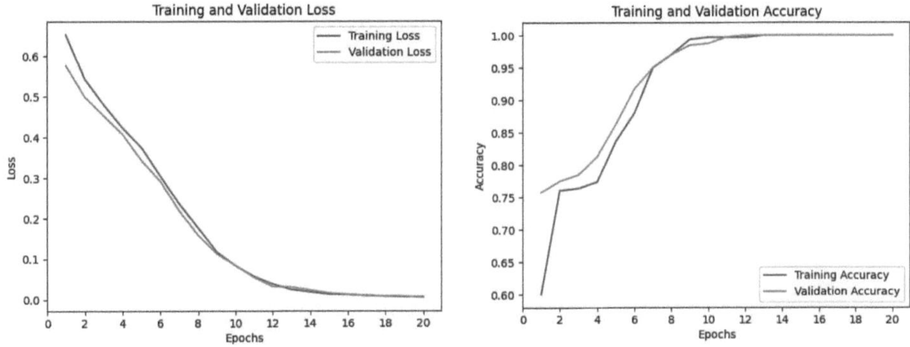

Fig. 3. Training and Validation Graphs of the experimental result

The experimental results help organizations make the proper decision to improve their employee retention rates. Compensation surely contributes to the retention of employees, but this is only a small part of a more complex picture. It is important, but they do not represent a loyal worker [2, 25]. Different factors, such as job satisfaction, work-life relations, and career opportunities, influence the employee's decision to remain in a company without being affected directly by money. Moreover, people's interest and respect towards the organization and communication with management and other members besides the employee also help retain the employee. Besides those, it is evident that the company's or organization's other characteristic features, such as culture, leadership, and benefits provided, should be considered as they can be helpful or harmful to retention.

7   Discussion and Conclusion

Through the use of artificial intelligence (AI) in human resource management (HRM) systems, transformation in companies' human resources has changed positively, enhanced decision-making, facilitated repetitive work, and offered data-driven insights, which has been revolutionary. This paper investigated the application of AI in managing human resources, explaining how it can influence management processes, raise performance, and support strategic decision-making. Despite these positive aspects, there are challenges, specifically ethical ones like bias, overuse of technology, and respect for human dignity, which must be handled with care before implementation. We also conducted a case study using AI to anticipate employee attrition and achieved remarkable accuracy, highlighting AI's revolutionary significance in HRM.

To summarize, it can be seen that there is a significant potential in utilizing AI for the efficiency of HRM. However, ethical concerns and operational problems also arise, which need attention. It is necessary to ensure all the measures that encourage trust in AI-enhanced HR functions, including the ethical use of the technology. In this regard, the focus for the future needs to be on creating responsible practices and adopting XAI, which will allow AI to be used in good ways. HR practitioners also have to advance their skills to exploit these as AI evolves.

References

1. Athanasios Chatzimouratidis, I.T., Lagoudis, I.N.: Decision support systems for human resource training and development. Int. J. Human Res. Manag. **23**(4), 662–693 (2012). https://doi.org/10.1080/09585192.2011.561235
2. Babatunde, F., Onoja, E.: The effectiveness of retention strategies on employee retention **13**, 1–5 (2023).https://doi.org/10.37421/2161-5833.2023.13.481
3. Bal, M., Bal, Y.: A novel approach for the recruitment process in human resources management: decision support system based on formal concept analysis. Int. J. Dec. Support Syst. Tech. **14**(1), 1–20 (2022). https://doi.org/10.4018/IJDSST.292447
4. Bankins, S.: The ethical use of artificial intelligence in human resource management: a decision-making framework. Ethics Inf. Technol. **23**(4), 841–854 (2021). https://doi.org/10.1007/s10676-021-09619-6
5. Bankins, S., Formosa, P., Griep, Y., Richards, D.: Ai decision making with dignity? contrasting workers' justice perceptions of human and ai decision making in a human resource management context. Inf. Syst. Front. **24**(3), 857–875 (2022). https://doi.org/10.1007/s10796-021-10223-8
6. Budhwar, P., et al.: Human resource management in the age of generative artificial intelligence: Perspectives and research directions on chatgpt. Human Res. Manag. J. **33**(3), 606–659 (2023).https://doi.org/10.1111/1748-8583.12524
7. Bujold, A., Roberge-Maltais, I., Parent-Rocheleau, X., Boasen, J., Sénécal, S., Léger, P.M.: Responsible artificial intelligence in human resources management: a review of the empirical literature. AI and Ethics (2023). https://doi.org/10.1007/s43681-023-00325-1
8. Cai, C., Chen, C.: Optimization of human resource file information decision support system based on cloud computing. Complexity **2021**(1), 8919625 (2021)
9. Chowdhury, S., et al.: Unlocking the value of artificial intelligence in human resource management through ai capability framework. Hum. Resour. Manag. Rev. **33**(1), 100899 (2023). https://doi.org/10.1016/j.hrmr.2022.100899, https://www.sciencedirect.com/science/article/pii/S1053482222000079
10. Czarnowski, I., Pszczółkowski, P.: A novel framework for decision support system in human resource management. Proc. Comput. Sci. **176**, 1548–1556 (2020).https://doi.org/10.1016/j.procs.2020.09.166, https://www.sciencedirect.com/science/article/pii/S1877050920320664, knowledge-Based and Intelligent

Information and Engineering Systems: Proceedings of the 24th International Conference KES2020

11. Dulebohn, J.H., Johnson, R.D.: Human resource metrics and decision support: a classification framework. Human Res. Manag. Rev. **23**(1), 71–83 (2013). https://doi.org/10.1016/j.hrmr.2012.06.005, https://www.sciencedirect.com/science/article/pii/S1053482212000484, emerging Issues in Theory and Research on Electronic Human Resource Management (eHRM)

12. Holzinger, A., Saranti, A., Molnar, C., Biecek, P., Samek, W.: Explainable AI Methods - A Brief Overview, pp. 13–38. Springer International Publishing, Cham (2022). https://doi.org/10.1007/978-3-031-04083-2_2, http://dx.doi.org/10.1007/978-3-031-04083-2_2

13. Hunkenschroer, A.L., Kriebitz, A.: Is ai recruiting (un)ethical? a human rights perspective on the use of ai for hiring. AI and Ethics **3**(1), 199–213 (2022). https://doi.org/10.1007/s43681-022-00166-4

14. Jantan, H., Hamdan, A.R., Othman, Z.A.: Potential intelligent techniques in human resource decision support system (hr dss). In: 2008 International Symposium on Information Technology. vol. 3, pp. 1–9 (2008).https://doi.org/10.1109/ITSIM.2008.4632047

15. Kaplan, D.: Structural equation modeling. In: Smelser, N.J., Baltes, P.B. (eds.) International Encyclopedia of the Social and Behavioral Sciences, pp. 15215–15222. Pergamon, Oxford (2001).https://doi.org/10.1016/B0-08-043076-7/00776-2, https://www.sciencedirect.com/science/article/pii/B0080430767007762

16. Kaur, M., Gandolfi, F.: Artificial intelligence in human resource management - challenges and future research recommendations. Rev. Int. Comparative Manag. **24**, 382–393 (2023).https://doi.org/10.24818/RMCI.2023.3.382

17. Kaur, M.: Exploring the use of chatbots and ai in human resource management: A focus on ChatGPT and its impact of ChatGPT on human resource management practices **4**, 2442–2445 (2023)

18. Köchling, A., Riazy, S., Wehner, M.C., Simbeck, K.: Highly accurate, but still discriminatory: a fairness evaluation of algorithmic video analysis in the recruitment context. Business Inf. Syst. Eng. **63**(1), 39–54 (2020). https://doi.org/10.1007/s12599-020-00673-w

19. Murugesan, U., Subramanian, P., Srivastava, S., Dwivedi, A.: A study of artificial intelligence impacts on human resource digitalization in industry 4.0. Dec. Anal. J. **7**, 100249 (2023).https://doi.org/10.1016/j.dajour.2023.100249, https://www.sciencedirect.com/science/article/pii/S2772662223000899

20. Nawaz, N., Arunachalam, H., Pathi, B.K., Gajenderan, V.: The adoption of artificial intelligence in human resources management practices. Int. J. Inf. Manag. Data Insights **4**(1), 100208 (2024). https://www.sciencedirect.com/science/article/pii/S266709682300054X

21. Samuel, O.W., Omisore, M.O., Atajeromavwo, E.J.: Online fuzzy based decision support system for human resource performance appraisal. Measurement **55**, 452–461 (2014). https://www.sciencedirect.com/science/article/pii/S0263224114002450

22. Shah, A.: Datasets in HR analytics applied ai. https://www.kaggle.com/datasets/aryashah2k/datasets-in-hr-analytics-applied-ai

23. Sithambaram, R.A., Tajudeen, F.P.: Impact of artificial intelligence in human resource management: a qualitative study in the malaysian context. Asia Pacific J. Human Res. **61**(4), 821–844 (2022). https://doi.org/10.1111/1744-7941.12356

24. Song, Y., Wu, R.: Analysing human-computer interaction behaviour in human resource management system based on artificial intelligence technology. Knowledge Management Research and Practice, pp. 1–10 (2021)
25. Sorn, M.K., Fienena, A.R.L., Ali, Y., Rafay, M., Fu, G.: The effectiveness of compensation in maintaining employee retention. OALib **10**(07), 1–14 (2023). https://doi.org/10.4236/oalib.1110394
26. Strohmeier, S., Piazza, F.: Artificial intelligence techniques in human resource management–a conceptual exploration. Intelligent Techniques in Engineering Management: Theory and Applications,D pp. 149–172 (2015)
27. Tambe, P., Cappelli, P., Yakubovich, V.: Artificial intelligence in human resources management: Challenges and a path forward. Calif. Manage. Rev. **61**(4), 15–42 (2019). https://doi.org/10.1177/0008125619867910
28. Varma, A., Dawkins, C., Chaudhuri, K.: Artificial intelligence and people management: a critical assessment through the ethical lens. Hum. Resour. Manag. Rev. **33**(1), 100923 (2023)
29. Votto, A., Liu, C.Z.: Transparent artificial intelligence and human resource management: a systematic literature review. In: Proceedings of the 56th Hawaii International Conference on System Sciences. HICSS, Hawaii International Conference on System Sciences (2023). https://doi.org/10.24251/hicss.2023.132, http://dx.doi.org/10.24251/HICSS.2023.132
30. Vrontis, D., et al.: Artificial intelligence, robotics, advanced technologies and human resource management: a systematic review. Int. J. Human Res. Manag. **33**, 1–30 (2021). https://doi.org/10.1080/09585192.2020.1871398

An Ontology-Based Personalized Expert System for Purple Star Astrology

Anh Chi Tuan, Khac Trung Nguyen, and Tran Ngoc Thang

Faculty of Mathematics and Informatics, Hanoi University of Science and Technology,
Hanoi, Vietnam
`thang.tranngoc@hust.edu.vn`

Abstract. Vietnamese Purple Star Astrology (VPSA) has long been revered as a potent tool for understanding destiny and unlocking the mysteries of individual lives. There have been some studies on astrology expert systems, but they are mainly based on classification algorithms and none for VPSA yet. This article is the first to propose a personalized expert system that utilizes ontology technology for VPSA. The system provides more profound and semantic insights by constructing a structured knowledge representation in an ontology that encapsulates the palaces and stars with their complex relationships. The expert system acts as an astrologer who can give predictions on various life aspects based on a specific VPSA chart. In this paper, we propose an architecture for a personalized VPSA expert system with knowledge representation, information extraction, and a reasoning machine. Our simulation example proves the benefits of an ontology-based approach, including the ability to systematically represent and reasoning capability.

Keywords: Purple Star Astrology · Expert System · Ontology · Knowledge Model · Ontology Reasoning

1 Introduction

Vietnamese Purple Star Astrology (VPSA), also known as Tử vi đẩu số, is a form of Chinese "Zi Wei Dou Shu" astrology that has been adapted and practiced in Vietnam (see [9]). It focuses on the movement and positions of stars to create a comprehensive life chart. By providing deep insights into a person's characteristics and potential life path, VPSA offers valuable guidance and predictions, blending ancient wisdom with cultural significance.

At the heart of VPSA are the concepts of the twelve palaces, the fourteen major stars with more than 100 different minor stars. The twelve palaces represent various facets of a person's life: Life Palace, Parents Palace, Karma Palace, Property Palace, Career Palace, Friends Palace, Travel Palace, Health Palace, Wealth Palace, Children Palace, Spouse Palace, and Siblings Palace. Each star

This work was funded by Rikkeisoft Corporation and supported by Center for Digital Technology and Economy (BK Fintech), Hanoi University of Science and Technology.

S. Thomassey et al. (Eds.): RAIDS 2024, LNICST 673, pp. 139–149, 2026.
https://doi.org/10.1007/978-3-032-14055-5_10

belongs to one and only palace. Astrologers analyze these elements' positions, interactions, and balance to provide a detailed interpretation of the individual's life path, potential challenges, and opportunities.

Recently, this traditional practice has been enhanced with technology, allowing for higher accuracy. There are many online platforms; however, they can only create a VPSA chart (sometimes inaccurate) and give simple, general explanations, not taking into account the complex relationships between stars and palaces in the VPSA chart. Therefore, this paper presents an innovative solution: an ontology-based personalized expert system for VPSA. Using ontologies, the system formalizes the domain knowledge of VPSA, including the complex relationships between different astrological elements and their interpretations. This approach provides a new way of VPSA representation and improves the precision of the personal readings. To our knowledge, this paper is the first to study ontology-based astrological representation.

The rest of this paper is organized as follows. Section 2 analyzes the background and related works. Section 3 presents our proposed ontology-based architecture. Sections 4 and 5 describe the knowledge base and ontology base of the system, respectively. The semantic reasoning is explained in Sect. 6. Section 7 is our simulation example, and Sect. 8 concludes with our main contributions.

2   Background and Related Works

2.1   Knowledge Representation

Knowledge representation in Web Ontology Language (OWL) is a crucial aspect of the Semantic Web, aiming to provide a structured and semantic way to represent complex information. Resource Description Framework (RDF) is a standard model for data interchange on the web that allows semantic representation and, thus, it is a foundation for ontologies [2]. RDF uses a simple structure to make statements about resources using subject-predicate-object expressions. OWL builds upon RDF and RDF Schema (RDFS) to formalize domain knowledge through classes, properties, individuals, and complex relationships. With OWL, it is possible to make the semantics of the information explicit, allowing machines to understand and process web content more intelligently. This enables more sophisticated querying, data integration, and decision-making processes, ultimately contributing to the development of intelligent applications that can understand and respond to complex queries and scenarios.

2.2   Query Language

The Simple Protocol and RDF Query Language (SPARQL) [7] is a powerful and flexible query language designed for querying and manipulating data stored in the RDF format. Since ontologies are structured representations of knowledge, defining concepts and their relationships, SPARQL can perform complex queries that search for specific patterns, classes, or properties within the ontology. It allows users to extract valuable insights by combining different patterns and performing reasoning tasks on the data.

2.3   Relevant Systems

Some researchers have the same interest in developing systems related to astrology. Kulkarni et al. converted astrological principles in horoscope prediction using data mining techniques [3]. They used Waikato Environment for Knowledge Analysis (WEKA), a collection of machine learning and data analysis, to analyze horoscopes in the database and predict the nature of a new horoscope. Tiwari and Barde also utilized WEKA to perform logistic regression and Naive Bayes to classify horoscopes and give astrological predictions for government jobs [8]. Recently, Paul developed a customized system to predict essential life events [5]. A database of rules for horoscope interpretation from ancient scripts was first built. Besides birth information, he also collected dates of significant events in a person's past (date of marriage, date of first employment, etc.) to give predictions. Finally, machine learning methods like logistic regression, Gaussian Naive Bayes, and support vector classification are applied to learn from users' data and feedback to appropriately match prediction rules.

Another approach is case-based reasoning, a problem-solving approach that leverages the knowledge of previously encountered, specific problem situations (cases) to address new problems. Rishi et al. utilized case-based reasoning for astrological predictions about professions [6]. Chaplot and his colleagues also applied case-based reasoning for predictions about famous personalities [1].

In general, existing systems usually use classification algorithms and can only answer a particular question about occupation or personality. On the other hand, some researchers adopted ontologies to represent complex relationships of entities. For example, Ntioudis et al. successfully built a job recommendation framework, which allows for the semantic representation and recommendations [4]. Therefore, our main idea is to develop an ontology-based expert system for VPSA. As far as we know, there is currently no ontology-based system that can provide astrological predictions on various aspects of life.

3   Proposed Architecture

This section provides an overview of the proposed VPSA expert system architecture. Figure 1 describes the high-level architecture of the framework, which shows how the different components are organized. The framework consists of many components, each with a specific function. These are the following: User Interface, VPSA Chart, Ontology base Repository, Knowledge base Repository, and Reasoning Machine.

The User Interface serves as the bridge between the user and the system. From VPSA documents, information is extracted and stored in the Knowledge Base Repository as either rules to create a VPSA chart or predictions. These rules are integrated manually to create a personalized VPSA chart. The chart is then expressed as an ontology and saved in the Ontology Base Repository. Finally, as predictions are If-Then conditional statements, the Reasoning Machine uses information from the ontology to check which predictions match the person's VPSA chart to make suggestions.

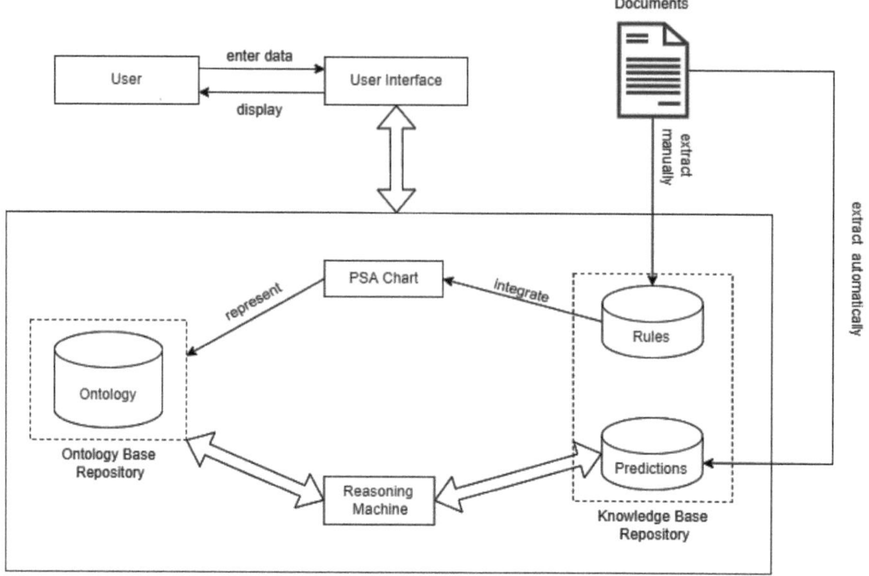

Fig. 1. Proposed architecture.

4   Knowledge Base

4.1   Rules to Create VPSA Chart

Based on information about the date and time of birth, we can appropriately arrange the positions of palaces, stars, and their characteristics. To be able to do that, an SQL database system is needed to store the characteristics of palaces and stars. Specifically, the information and rules for determining the location of objects (palaces, stars) are manually extracted from VPSA books and integrated into the suggested system. Through these rules and information, along with a user's input data, we can create a personalized VPSA chart automatically.

4.2   Knowledge for Intepretation

Figure 2 describes the information extraction process from books and online resources. Firstly, ancient VPSA documents from books and online resources are converted into PDF/txt files. Then, horoscope aphorisms, prediction sentences, and references are extracted. Typically, prediction sentences are IF <conditions> THEN <results> statements; thus, we will continue to identify conditions and results of statements. As conditions might contain different objects with complex relationships and prediction sentences often follow a consistent and predictable format, regular expressions are applied to find entities and relations.

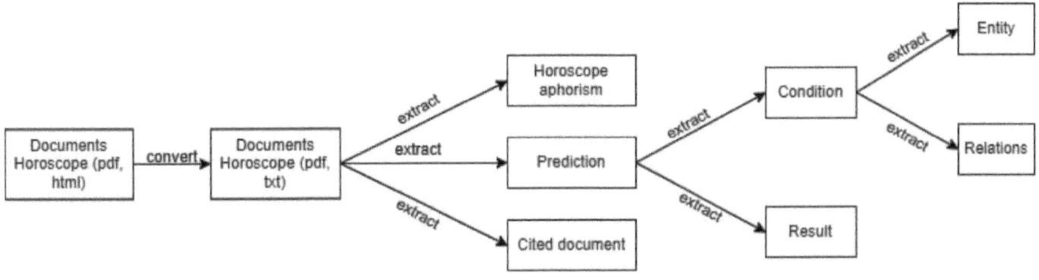

Fig. 2. Prediction extraction process.

5   Ontology

A VPSA ontology is defined as the following tuple:

$$<C, I, R, O, D, A>, \tag{1}$$

in which

- C: Set of classes.
- I: Set of instances or individuals.
- R: Relation between two classes (subclasses versus superclasses) or between an instance and a class.
- O: Object property (relation) between two instances.
- D: Data property (relation) between an instance and a data value with its type.
- A: Annotation property (relation) between an object (class, instance, object/data property, even ontology itself) and a string.

5.1   Representing Personal Information

The representation of the personal information of a specific user in the ontology is illustrated in Fig. 3. At the center is the class Person, which has four individuals: Lunar Calendar, Solar Calendar, Year of Prediction, and Other. Each individual is associated with a set of data properties and an annotation property, as shown in the figure. The individual's label is the individual's official English name and differs from others. These labels can be utilized instead of class/individual indices during the querying phase later.

5.2   Representing Palaces

In Fig. 4, the class Palace consists of 12 palaces: Life Palace, Parents Palace, Karma Palace, Property Palace, Career Palace, Friends Palace, Travel Palace, Health Palace, Wealth Palace, Children Palace, Spouse Palace, and Siblings Palace. Each palace has different characteristics and corresponding labels, as depicted by its many data and annotation properties. Palaces are linked via

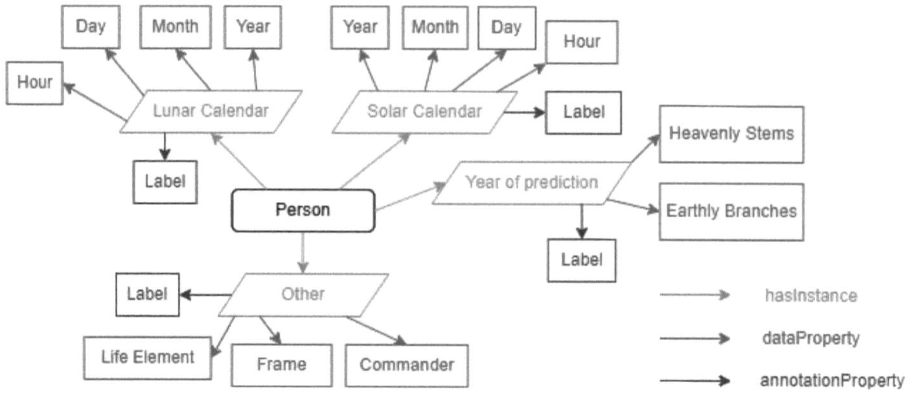

Fig. 3. Personal information representation.

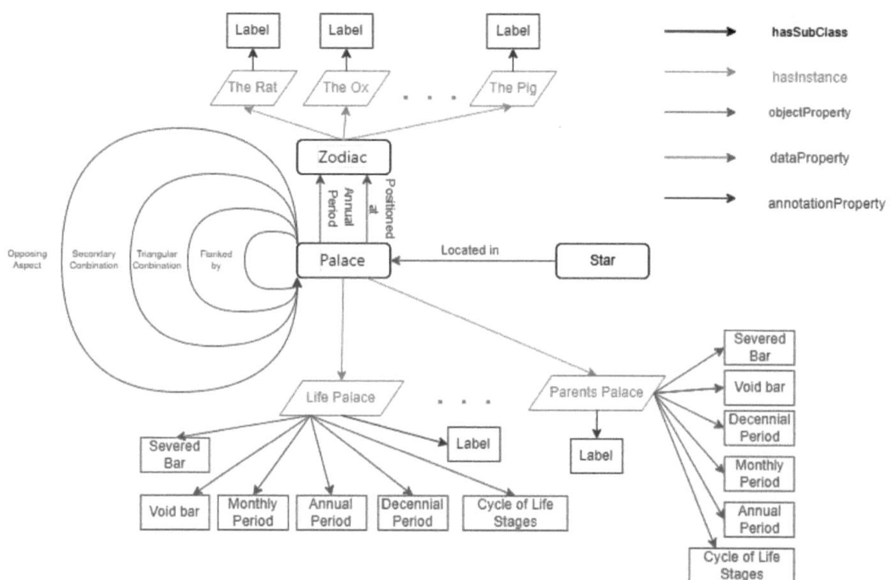

Fig. 4. Palace representation.

object properties with various relationships. Each star in the class Star lies on only one palace. The Vietnamese zodiac system also plays an important role in Vietnamese astrology. Twelve individuals in the class Zodiac are Rat, Ox, Tiger, Cat, Dragon, Snake, Horse, Goat, Monkey, Rooster, Dog, and Pig. This interconnected structure generally provides a comprehensive and accurate representation of the complex relationships of the class Palace.

5.3   Representing Stars

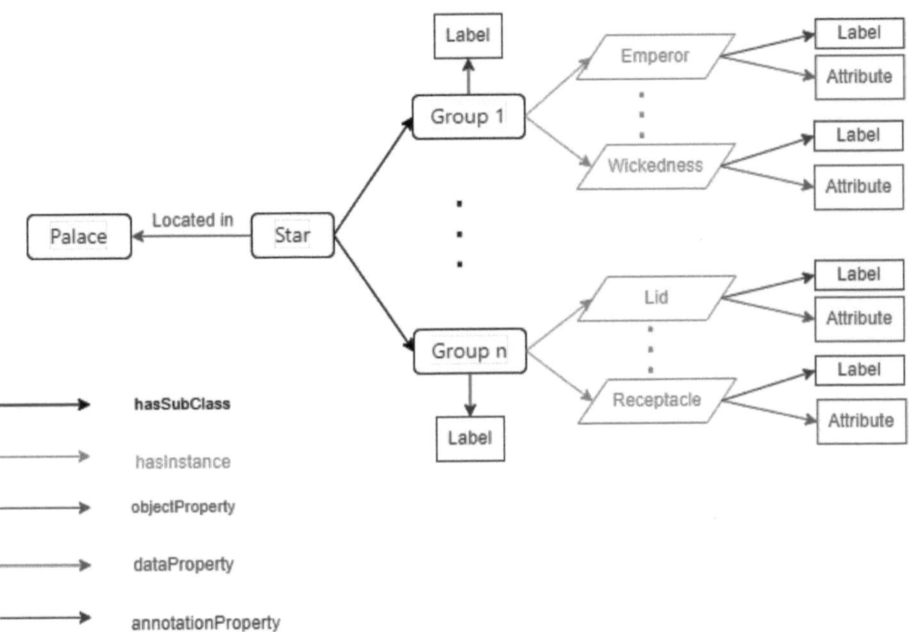

Fig. 5. Star representation.

Figure 5 is the representation of stars in the ontology for the VPSA system. Each star has its label and attribute. A star attribute shows its strong or weak, positive or negative influence on the person. Stars are categorized into distinct groups based on their unique characteristics, functions, and significance in analyzing and predicting human fate. Grouping facilitates astrologers' ability to identify and interpret factors affecting an individual's life. Finally, each star in the class Star belongs to a specific palace. Depending on the characteristics of each star, it can stay in a specific palace forever or relocate to a different palace when times change.

6   Semantic Reasoning

6.1   Reasoning Framework

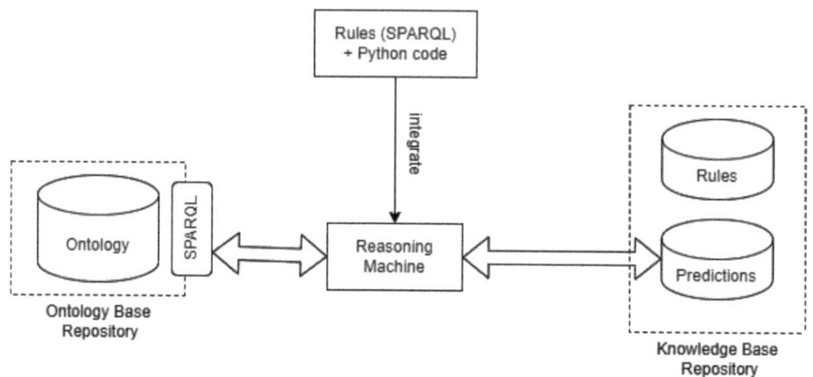

Fig. 6. Reasoning framework.

The reasoning framework for the VPSA expert system, as depicted in Fig. 6, outlines a structured approach to integrating an ontology base with a knowledge base repository with inferential reasoning to provide personalized astrological insights. The Ontology Base Repository stores the corresponding ontology representation of a VPSA chart. The database Predictions contains different astrological predictions extracted and pre-processed from various ancient VPSA books, scripts, or online sources. Rules are written in SPARQL and combined with Python codes to input into the Reasoning Machine. The Reasoning Machine will infer new knowledge, check for accurate conditions in If-Then statements from the database Predictions, and then give suitable astrological predictions. Specifically, if the hypothesis of an if-then statement is true, its conclusion will be stored and displayed.

7   Simulation Example

7.1   Initialization Step

Figure 7 describes the input section for a user to enter the information, which includes Name, Gender, Date of Birth, Hour of Birth, Year of Prediction, and References source of VPSA documents.

7.2   Results

The person's VPSA chart is stored as an ontology. Figure 8 describes the graph of the corresponding VPSA ontology in Fig. 9. Note that data properties and annotation properties are not shown in this figure. Based on the ontology, relevant

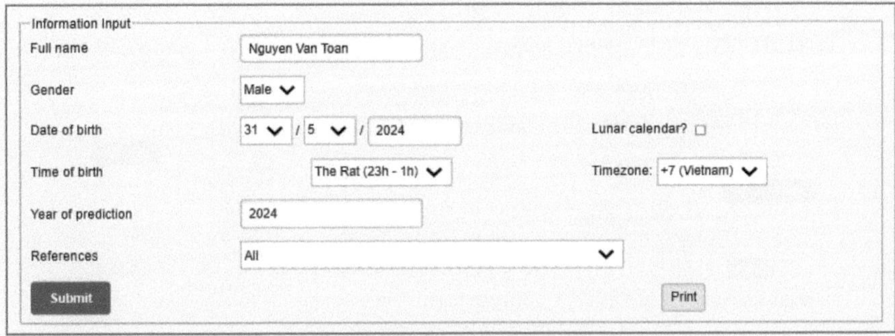

Fig. 7. Input section in the user interface.

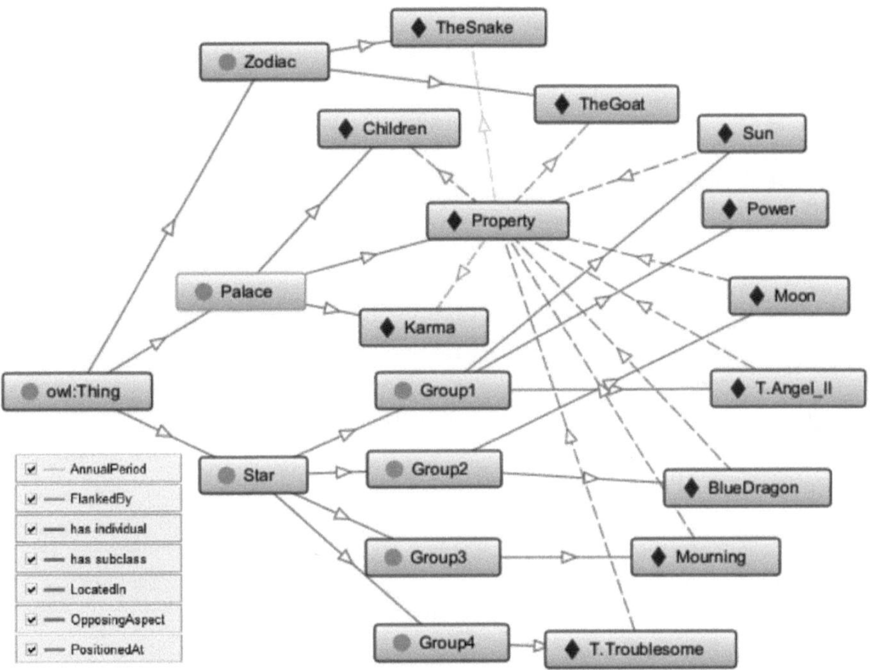

Fig. 8. Ontology graph.

predictions are found and displayed through the reasoning machine, for example, "If Property palace has stars Power and Prosperity encountering Emperor, then he is a wealthy person".

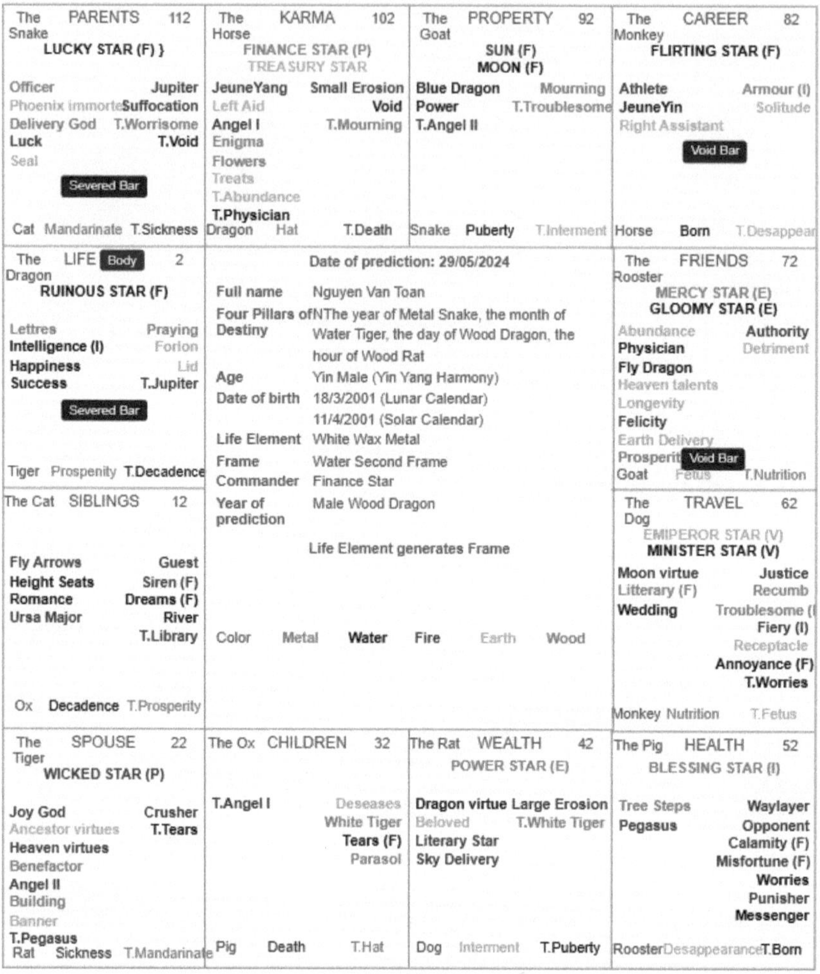

The Snake — PARENTS — 112
LUCKY STAR (F) }
Officer — Jupiter
Phoenix immortel — Suffocation
Delivery God — T.Worrisome
Luck — T.Void
Seal
[Severed Bar]
Cat — Mandarinate — T.Sickness

The Horse — KARMA — 102
FINANCE STAR (P)
TREASURY STAR
JeuneYang — Small Erosion
Left Aid — Void
Angel I — T.Mourning
Enigma
Flowers
Treats
T.Abundance
T.Physician
Dragon — Hat — T.Death

The Goat — PROPERTY — 92
SUN (F)
MOON (F)
Blue Dragon — Mourning
Power — T.Troublesome
T.Angel II
Snake — Puberty — T.Interment

The Monkey — CAREER — 82
FLIRTING STAR (F)
Athlete — Armour (I)
JeuneYin — Solitude
Right Assistant
[Void Bar]
Horse — Born — T.Desappear

The Dragon — LIFE [Body] — 2
RUINOUS STAR (F)
Lettres — Praying
Intelligence (I) — Forlon
Happiness — Lid
Success — T.Jupiter
[Severed Bar]
Tiger — Prospenity — T.Decadence

The Cat — SIBLINGS — 12
Fly Arrows — Guest
Height Seats — Siren (F)
Romance — Dreams (F)
Ursa Major — River
T.Library
Ox — Decadence — T.Prosperity

Date of prediction: 29/05/2024

Full name	Nguyen Van Toan
Four Pillars of Destiny	NThe year of Metal Snake, the month of Water Tiger, the day of Wood Dragon, the hour of Wood Rat
Age	Yin Male (Yin Yang Harmony)
Date of birth	18/3/2001 (Lunar Calendar)
	11/4/2001 (Solar Calendar)
Life Element	White Wax Metal
Frame	Water Second Frame
Commander	Finance Star
Year of prediction	Male Wood Dragon

Life Element generates Frame

Color	Metal	**Water**	**Fire**	Earth	Wood

The Rooster — FRIENDS — 72
MERCY STAR (E)
GLOOMY STAR (E)
Abundance — Authority
Physician — Detriment
Fly Dragon
Heaven talents
Longevity
Felicity
Earth Delivery
Prosperity [Void Bar]
Goat — Fetus — T.Nutrition

The Dog — TRAVEL — 62
EMIPEROR STAR (V)
MINISTER STAR (V)
Moon virtue — Justice
Litterary (F) — Recumb
Wedding — Troublesome (I)
— Fiery (I)
— Receptacle
— Annoyance (F)
— T.Worries
Monkey — Nutrition — T.Fetus

The Tiger — SPOUSE — 22
WICKED STAR (P)
Joy God — Crusher
Ancestor virtues — T.Tears
Heaven virtues
Benefactor
Angel II
Building
Banner
T.Pegasus
Rat — Sickness — T.Mandarinate

The Ox — CHILDREN — 32
T.Angel I — Deseases
— White Tiger
— Tears (F)
— Parasol
Pig — Death — T.Hat

The Rat — WEALTH — 42
POWER STAR (E)
Dragon virtue — Large Erosion
Beloved — T.White Tiger
Literary Star
Sky Delivery
Dog — Interment — T.Puberty

The Pig — HEALTH — 52
BLESSING STAR (I)
Tree Steps — Waylayer
Pegasus — Opponent
— Calamity (F)
— Misfortune (F)
— Worries
— Punisher
— Messenger
Rooster — Desappearance — T.Born

Fig. 9. A person's VPSA chart.

8 Conclusion

8.1 Main Contributions

The article's main contribution is the development of an ontology-based expert system for VPSA, offering a structured and semantic approach to astrological interpretation. This expert system leverages a comprehensive ontology to encode the rich and complex knowledge inherent in Vietnamese astrology. By integrating this ontology with inferential reasoning using SPARQL queries and Python code, the system can deliver highly personalized astrological readings. It enhances traditional practices by providing a systematic, scalable, and precise method for astrological analysis. The framework's detailed representation of temporal data and astrological elements ensures accurate and meaningful predictions, making a significant advancement in the digital application of VPSA.

8.2 Future Works

Future work for an ontology-based astrology expert system could involve several key areas, such as (i) Revising and improving the ontology representation to represent knowledge better and allow for more accurate predictions, (ii) Creating a mechanism for users to provide feedback on the accuracy and relevance of the predictions to improve accuracy, (iii) Implementing natural language processing techniques and large language models to create a chatbot which allows users to ask questions and receive responses in natural language.

References

1. Chaplot, N., Dhyani, P., Rishi, O.P.: Predictive approach of case base reasoning in artificial intelligence: in case of astrological predictions about famous personalities. In: Proceedings of the Second International Conference on Information and Communication Technology for Competitive Strategies, ICTCS 2016. Association for Computing Machinery, New York (2016). https://doi.org/10.1145/2905055.2905148
2. Decker, S., et al.: The semantic web: the roles of XML and RDF. IEEE Internet Comput. **4**(5), 63–73 (2000). https://doi.org/10.1109/4236.877487
3. Kulkarni, P., Sane, D., Bhale, N.: Use of neural networks in horoscope prediction (2012). https://doi.org/10.13140/2.1.2394.2403
4. Ntioudis, D., Masa, P., Karakostas, A., Meditskos, G., Vrochidis, S., Kompatsiaris, I.: Ontology-based personalized job recommendation framework for migrants and refugees. Big Data Cogn. Comput. **6**(4) (2022). https://doi.org/10.3390/bdcc6040120
5. Paul, P.S.: Application of big data and machine learning for astrological predictions. In: Computational Intelligence in Pattern Recognition, pp. 1–12. Springer, Singapore (2022)
6. Rishi, O., Chaplot, N.: Predictive role of case based reasoning for astrological predictions about profession: system modeling approach. In: 2010 International Conference on Communication and Computational Intelligence (INCOCCI), pp. 313–317 (2010)
7. Harris, S., Seaborne, A., Prud'hommeaux, E.: Sparql 1.1 query language. In: W3C Recommendation (2013)
8. Tiwari, S., Barde, S.: Machine learning make possible to astrological prediction for government job using classification techniques. In: 2021 Asian Conference on Innovation in Technology (ASIANCON), pp. 1–4 (2021). https://doi.org/10.1109/ASIANCON51346.2021.9544586
9. Tu, V.V.: The Mysteries of Purple Star Astrology (Vietnamese). Saigon Publishing House (1972)

Aspect and Sentiment Detection in Vietnamese Text Using Transfer Learning and LSTM

Van Uc Ngo[1]([⊠]) [iD], Dat Vo Ngoc[2] [iD], Quan Ngo Le[2] [iD], Thin Nguyen Si[2] [iD], and Van Quan Pham[1] [iD]

[1] Dong A University, Da Nang, Vietnam
`{ucnv,quan52645}@donga.edu.vn`
[2] Vietnam - Korea University of Information and Communication Technology - The University of Da Nang, Da Nang, Vietnam
`{vndat,nlquan,nsthin}@vku.udn.vn`

Abstract. Aspect detection and sentiment classification is a challenging task in the field of natural language processing (NLP), especially in Vietnamese due to its complex grammatical structure and diversity of expressions. In this study, we propose a method of using transfer learning (BERT) and LSTM to simultaneously detect aspects and sentiments on customer feedback on mobile phone products. BERT is used to extract semantic features from text, while LSTM processes these features to aspects detection and sentiment classification. These results show that our method outperforms traditional deep learning methods. The experiment results show that our model achieved an accuracy of 93.5%, F1-score of 88.7% on the overall performance of the model. For the aspects detections, our model get accuracy at 81.5% and F1-score at 89.9%. And for the sentiment classification, an accuracy of 90.6% and F1-score at 90.6% on all aspects of model. This paper compares the performance of BERT combined with RNN, GRU, and Transformer models, provides detailed evaluations of their effectiveness, discusses their limitations, and suggests future research directions to further enhance their practical applications.

Keywords: Transfer learning · BERT · LSTM · Natural Language Processing · Aspect Based Sentiment Analysis

1 Introduction

The growth of e-commerce has changed the way consumers make purchasing decisions, in which they rely more and more on the feedback and reviews of previous users to judge the quality of products. These responses not only help consumers make decisions, but they also provide merchants with valuable information to tailor products and services, improve customer experience, and build effective business strategies.

Stemming from the above problems, the problem of detecting aspects and emotions was born as a field in Artificial Intelligence (AI) and natural language processing (NLP)

© ICST Institute for Computer Sciences, Social Informatics and Telecommunications Engineering 2026
Published by Springer Nature Switzerland AG 2026. All Rights Reserved
S. Thomassey et al. (Eds.): RAIDS 2024, LNICST 673, pp. 150–161, 2026.
https://doi.org/10.1007/978-3-032-14055-5_11

that aims to automatically determine the sentiment of a text, usually according to categories such as positive, negative or neutral in a certain aspect [1]. However, applying this aspect and sentiment detection problem to business analysis to understand user feedback is still difficult, especially for Vietnamese text, which is a language with a complex grammatical structure and sometimes implicit usage.

Table 1 below shows examples of Vietnamese user feedback translated into English, with the paragraphs highlighted in the same color. These examples were taken from the dataset used in this study.

Table 1. Three examples from the customer feedback dataset on mobile phone products.

Example Comment Sentences	Aspect of Sentiment Label
"Just bought this device at Thegioididong Thot Not feel it's okay. The battery is strong {BATTERY#Positive}, takes good photos {CAMERA#Positive} loudspeakers, strong Wi-Fi connection {FEATURES#Positive}, stable signal, reasonably priced {PRICE#Positive}, and the staff are very enthusiastic in advising {SER&ACC#Positive}." {GENERAL#Positive} "Mới mua máy này Tại thegioididong thốt nốt cảm thấy ok bin trâu {BATTERY#Positive}, chụp ảnh đẹp {CAMERA#Positive}, loa nghe to bắt wf khỏe sóng ổn định {FEATURES#Positive}, giá thành vừa với túi tiền {PRICE#Positive}, nhân viên tư vấn nhiệt tình {SER&ACC#Positive}." {GENERAL#Positive}	{CAMERA#Positive}; {FEATURES#Positive}; {BATTERY#Positive}; {PRICE#Positive}; {GENERAL#Positive}; {SER&ACC#Positive}
"The battery is poor {BATTERY#Negative}, but everything else is great {GENERAL#Positive}. Bought it on 8/3/2019 and the battery health is down to 88%. Is anyone else experiencing this? {OTHERS}" "Pin kém {BATTERY#Negative} còn lại miễn chê {GENERAL#Positive} mua 8/3/2019 tình trạng pin còn 88% có ai giống tôi không {OTHERS}"	{BATTERY#Negative}; {GENERAL#Positive}; {OTHERS};
"Everyone, update your software, it will help reduce battery drain {BATTERY#Neutral}. I've tried it, and everything is okay, but the fingerprint sensor isn't very responsive {FEATURES#Negative}". {GENERAL#Neutral} "Mọi người cập nhật phần mềm lại , nó sẽ bớt tốn pin {BATTERY#Neutral}, mình đã thử rồi, mọi thứ cũng ok, nhưng vân tay ko nhạy {FEATURES#Negative}". {GENERAL#Neutral}	{FEATURES#Negative}; {BATTERY#Neutral}; {GENERAL#Neutral};

Although many methods have been proposed to solve this problem, previous studies have solved this problem by separating it into two small tasks: extracting opinion targets and detecting emotional polarization [2]. Traditional and modern machine learning methods have also been used to extract objective views and detect emotions simultaneously [2]. Each method has certain advantages. However, these methods are still not highly accurate and need further improvement.

Recently, the development of large language models (LLMs) has opened up new opportunities in improving the accuracy and efficiency of the problem of detecting aspects and emotions in Vietnamese texts. LLM models such as BERT, GPT and other variants have demonstrated a high level of text understanding and semantic processing [3] to improve the quality of text analysis.

Therefore, in this paper, we propose a method that combines transfer learning with BERT and LSTM models to simultaneously detect aspects and classify sentiment on the Vietnamese dataset of users' responses to the phone, in order to take advantage of the power of large language models and improve the prediction accuracy on the Vietnamese documents.

With the advantage of reducing training costs and processing time when performing detection and classification for input. This method is aimed at practical application in e-commerce systems, customer experience management, and big data analytics platforms. This method has the potential for wide application not only in e-commerce, but also in other fields such as customer experience management and big data analytics. Our research results provide an effective solution for Vietnamese text processing and improve the quality of user feedback analysis.

Key contributions to this article include:

- Propose a combined method of BERT and LSTM for aspect detection and sentiment classification.
- Using transfer learning from large language models to improve the accuracy and efficiency of Vietnamese text analysis.
- Provides a comparative of the combination of BERT with RNN, GRU, and Transformer models for aspect detection and sentiment classification tasks. The evaluation is conducted to objectively assess the performance of these models, offering insights into their effectiveness and applicability within the domain.
- Comprehensively evaluate the effectiveness of the model on multiple aspects and discuss limitations and propose future improvement directions.

2 Related Work

Aspect-based sentiment analysis has become an important area of study in recent years, especially with the strong development of deep learning techniques and large language models. A variety of methods have been proposed to address related challenges, from traditional machine learning models to modern deep learning methods, which improve performance in aspects of detection and sentiment classification.

Luc et al. (2021) studied the use of BiLSTM (Bidirectional Long Short-Term Memory) network for aspect-based sentiment analysis, but the results were not positive when the accuracy was only average, especially for Vietnamese texts [4]. This study shows that traditional deep learning models have difficulty in handling the complex and multi-meaning contexts that are typical of the Vietnamese language.

Thin et al. (2018) extended the above method by using neural networks to detect aspects for Vietnamese, thereby identifying pairs of entities and attributes expressed in the text [6]. And TN Si et al. (2021) in Vietnamese sentiment analysis problem, the study indicated that restricted by non-expansion feature of data input [10]. This research proposed a featured combine model by integrating matrix factorization in collaborative filtering and sentiment analysis in text mining to predict rating user. Although the results are improved compared to traditional methods, the accuracy is still significantly lower. One of the main problems is the limitation of these methods in processing large texts and the semantic flexibility of the Vietnamese.

The advent of large language models (LLMs) such as BERT and GPT marked a major step forward in the field of natural language processing, especially in context-based and semantic text analysis. These LLM models represent a major step forward in Artificial Intelligence (AI) and especially towards the goal of human-like synthetic artificial intelligence [5]. BERT (Bidirectional Encoder Representations from Transformers), developed by Devlin et al. (2019) has demonstrated a strong ability to process the two-dimensional semantics of text and understand context more deeply than previous models [8]. Studies such as those of Mickel et al. (2019) have applied BERT to face-based sentiment analysis and have shown results that are superior to traditional deep learning methods [1].

Thin et al. (2022) proposed an effective generic multitasking architecture based on neural network models to solve two tasks in ABSA, which is designed to predict the entire category of aspects in question and emotional polarizations [7]. The results of this study show that the model achieves high performance in analyzing sentiment based on the aspect but has not actually achieved the expected results.

Although BERT has achieved great success in many languages, the application of BERT to Vietnamese is still limited due to the lack of large-scale training data and specialized models for the language. To overcome this, many recent studies have combined BERT with other deep learning models, such as LSTM, to harness the power of both models in aspect detection and sentiment classification on Vietnamese texts [7]. This combined method improves accuracy by taking advantage of BERT's text-specific extraction capabilities and LSTM's data string processing capabilities.

Our research continues this development direction, focusing on the use of the BERT large language model in combination with LSTM to simultaneously solve both the task of aspect detection and sentimental classification on Vietnamese texts. Not only do we improve performance compared to traditional methods, but we also emphasize the importance of using transfer learning to enhance the generalization capabilities of the model.

3 Method Propose

In this study, we used a dataset of customer feedback on mobile phone products in Vietnamese, we built a comprehensive process to achieve the goal in this study. We first selected BERT. The dataset and BERT version we use are both open and free for research purposes.

Next, we use BERT to convert sentences into a format that BERT can handle. We decided to keep the sentence format unchanged without any pure text processing, in order not to lose the style of the semantic features, thereby helping the model to learn the most complex features in Vietnamese.

In each input sentence, we stipulate a maximum length of 128 words. Sentences of greater length will be truncated, while shorter sentences will be processed using the Padding technique to ensure all sentences are 128 tokens in length. Adding Padding tokens has no semantic meaning and will not affect the learning process of the model [8]. Adjusting the maximum length makes the model work better with long sentences but also increases the computational cost.

We then encode the sentences into tokens to train BERT. We use a multilingual version of *bert-base-multilingual-cased*. After BERT extracts the features from the text and converts them into 768-dimensional characteristic vectors, we build the LSTM architecture to process the output of the BERT. The GRU, RNN and LSTM architecture is set up with 256 hidden units and processed in 2 directions of the chain before giving the result. We add a linear layer to perform the classification of the model's outputs with the number of outputs calculated using the formula: *Number of output* $= n_{classes} * n_{label\_per\_class}$. Where $n_{classes}$ is the number of aspects and $n_{label\_per\_class}$ is the number of emoticons for each aspect. With Transformer, following BERT a Transformer encoder is built with 6 layers, each having 8 attention heads and a feedforward dimension of 2048. This architecture processes the BERT outputs to capture deeper contextual information. Finally, a linear layer is applied for classification, with the number of outputs calculated as *Number of output*. These mean that the models make predictions for multiple layers and multiple labels on each layer. Note that in this study we specified the output label as "None" for classes that are not in the text.

3.1 Dataset

The UIT-ViSFD dataset used in this study has been previously described in detail by Luc et al. (2021) [4]. Below, we provide a summary of the dataset's key characteristics relevant to this study.

This dataset was collected from a Vietnamese e-commerce website. To ensure diversity, objectivity, and accuracy in the data set, responses from 10 phone brands were collected. The annotation principle in a dataset is followed by a strict annotation process to ensure data quality. The data was divided into three separate volumes: the training (Train), development (Dev) and testing (Test) with a ratio of 7:1:2 respectively.

Table 2. Overview statistics of Train/Dev/Test sets of UIT-ViSFD dataset [4].

	Train	Dev	Test
Number of Comments	7,786	1,112	2,224
Number of Tokens	283,460	39,023	80,787
Number of Aspects	23,597	3,371	6,742
Average number of aspects per sentence	3.3	3.2	3.3
Average length per sentence	36.4	35.1	36.3

Table 2 presents the overview statistics of the UIT-ViSFD dataset. Number of Comments is the number of comments collected, Number of Tokens is the number of words, Number of Aspects is the number of aspects extracted from comments, Average number of aspects per sentence is the number of aspects per sentence and Average length per sentence is the average length of sentences.

Table 3 depicts the distribution of aspects and sentiments in the Train, Dev, and Test suites. The aspects have a marked imbalance, the most noticeable being that of the GENERAL class is 6,936 data points while STORAGE has only 132. There is also a big difference between the three emotional poles. The number of sensory counts also varies greatly, with Pos accounting for 56.13%, followed by Neg at 31.70% and finally Neu at 12.17%. This dataset is unbalanced [4].

Table 3. The distribution of aspects and their sentiments of UIT-ViSFD dataset [4].

Aspect	Train			Dev			Test			Total
	Pos	Neg	Neu	Pos	Neg	Neu	Pos	Neg	Neu	
BATTERY	2,027	349	1,228	303	51	150	554	92	368	5,122
CAMERA	1,231	288	627	172	36	88	346	71	171	3,030
DESIGN	999	77	302	135	12	40	274	28	96	1,963
FEATURES	785	198	1,659	115	33	233	200	52	459	3,734
GENERAL	3,627	290	949	528	34	127	1,004	83	294	6,936
PERFORMANCE	2,253	391	1,496	327	45	210	602	116	454	5,894
PRICE	609	391	316	72	144	36	162	328	79	2,882
SCREEN	514	56	379	62	12	47	136	17	116	1,339
SER&ACC	1,401	107	487	199	13	78	199	27	167	2,678
STORAGE	59	107	21	11	1	2	18	3	6	132
Total	13,505	2,903	7,464	1,924	381	1,011	3,495	817	2,210	

This dataset presents several challenges for aspect detection and sentiment classification due to the significant class imbalance. The imbalance of data can lead to two problems, the first is that the model will learn better in large numbers of classes, but fewer classes will tend to be ignored. The second is that the model may have more difficulty in learning the characteristics of the STORAGE class, leading to the model being mispredicted for classes with too few labels, thereby making the recognition less accurate. For instance, the STORAGE aspect has only 132 instances, compared to 6,936 instances for the GENERAL aspect. Additionally, many comments contain multiple aspects in the same sentence, making it harder for the model to accurately detect and classify sentiments for each aspect.

3.2 State-Of-The-Art Models

In this study, our overview model is built on a combination of two advanced models, BERT and LSTM, for both aspect detection and sentiment classification tasks simultaneously for a single output.

BERT architecture consists of a multi-layered two-dimensional Transformer encoder based on the original implementation described and released in the *tensor2tensor* library. The baseline BERT model we used in this study had the following parameters: $L = 12$, $H = 768$, $A = 12$, and the total number of parameters was 110M. Where L is the number of layers, the hidden layer size is H and A is the number of self-attention heads [8]. BERT is a powerful linguistic model that uses an attention mechanism to learn the semantic features of text in two dimensions. The BERT model is trained in large amounts of data and can capture deep semantic relationships in text thanks to the use of a two-dimensional self-attention mechanism [8]. The BERT model works according to the "*Masked Language Model (MLM)*" mechanism during the training phase, in which some words in the sentence are obscured, and the model must predict these obscured words based on the surrounding context. In addition, BERT is also trained with the "*Next Sentence Prediction (NSP)*" mechanism to understand the relationship between sentences in the text [8]. These characteristics allow BERT to provide very powerful semantic characteristic vectors, which can be used in a variety of natural language processing tasks, including sentiment analysis, information extraction, and many others.

RNN Architecture: RNNs are designed to process sequential data by maintaining a hidden state h_t that depends on both the current input x_t and the previous hidden state h_{t-1}. The update for the hidden state is given by the formula [11]:

$$h_t = \tanh(W_{xh}x_t + W_{hh}h_{t-1} + b_h)$$

RNNs are capable of handling sequential information but often struggle with the vanishing gradient problem.

GRU Architecture: The GRU is a variant of RNN that mitigates the vanishing gradient issue by introducing gating mechanisms. It uses two main gates: the update gate and the reset gate, to control the flow of information [12]:

Update gate:

$$z_t = \sigma(W_{xz}x_t + W_{hz}h_{t-1})$$

Reset gate:

$$r_t = \sigma(W_{xr}x_t + W_{hr}h_{t-1})$$

The new hidden state is computed as follows:

$$h_t = (1 - z_t) \cdot h_{t-1} + z_t \cdot \tanh(W_{xh}x_t + r_t \cdot W_{hh}h_{t-1})$$

GRUs help retain relevant information over longer time periods and efficiently update the hidden state.

LSTM architecture was first introduced by S Hochreiter el al. [9] is designed to process and predict sequential data, which is especially useful when data has a long-term dependency such as text. LSTM has a special structure with the following main steps:

This Gate: decides how much information from the previous state C_{t-1} will be retained. The Forget Gate receives input from the hidden state of the previous step h_{t-1} and the current input x_t then applies a sigmoid function to calculate the value f_t:

$$f_t = \sigma\left(W_f \cdot [h_{t-1}, x_t] + b_f\right)$$

The value f_t is in the range [0,1], where 0 means completely forget and 1 means fully retain.

Input Gate: This gate decides what new information will be added to the Cell State C_t. There are two main parts: First, a sigmoid function decides which values need to be updated, resulting in i_t:

$$i_t = \sigma\big(W_i \cdot \big[h_{t-1}, x_t\big] + b_i\big)$$

Then, a *tanh* layer creates a new vector of values $\tilde{C}_t$:

$$\tilde{C}_t = tanh\big(W_C \cdot \big[h_{t-1}, x_t\big] + b_C\big)$$

Update Cell State: The Cell State is updated by combining the information that needs to be forgotten and the new information:

$$C_t = f_t \cdot C_{t-1} + i_t \cdot \tilde{C}_t$$

Here, the multiplication $\cdot$ is element-wise multiplication.

Output Gate: Finally, to determine the output for the current hidden state h_t, the LSTM uses the output gate. First, a sigmoid function calculates which part of the Cell State will be output as o_t:

$$o_t = \sigma\big(W_o \cdot \big[h_{t-1}, x_t\big] + b_o\big)$$

Then, the current hidden state h_t is calculated by multiplying o_t by the tanh of the new Cell State C_t:

$$h_t = o_t \cdot \tanh(C_t)$$

Transformer Model: relies on the attention mechanism, eliminating recurrence to process data in parallel. The core component is Multi-head Self-Attention, which calculates interactions between words in a sequence. Given Q (query), K (key), and V (value), the attention scores are computed as [13]:

$$Attention(Q, K, V) = softmax(\frac{QK^T}{\sqrt{d_k}})V$$

The Transformer consists of multiple encoder and decoder layers, each comprising a self-attention mechanism and feed-forward network. It excels at modeling long-range dependencies and efficiently handles long sequences via its attention mechanism.

3.3 Our Model Performance Evaluation Method

In this study, we evaluate the model from various dimensions to examine its performance from a holistic to a detailed perspective. First, we assess the overall performance of the model to simultaneously check its ability to detect and classify aspects and sentiments. Next, we evaluate the model's performance in detecting aspects for each individual aspect as well as across all aspects. Similarly, we assess the model's sentiment classification performance for each aspect and across all aspects. A detailed evaluation from various dimensions is necessary to gain a deeper understanding of the model's capabilities, as well as to identify its strengths and limitations in specific aspects. The performance metrics used in the evaluation include Accuracy, Precision, Recall, and F1-score, which help measure the model's ability to detect and classify sentiments.

4 Experimental Results

In this study, we neglected to detect the aspect and categorize the emotions for the OTHERS class because they do not express their own emotions in the paragraph. Figure 1 shows the models' loss index during training starting at 0.35 and converging at 0.01 after going through 40 epochs. The descending loss graph shows that the model has learned well-learned features that can distinguish more accurately over multiple epochs. All models converge best after 40 epochs and have approximately equal values.

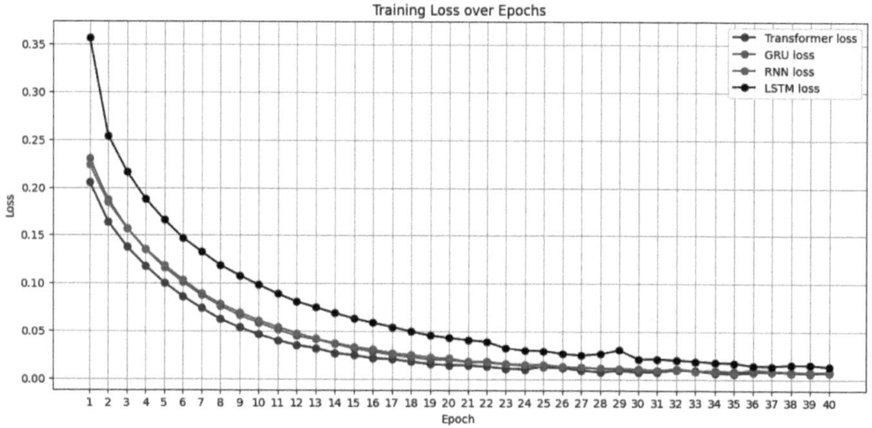

Fig. 1. The loss metric chart of the models over 40 epochs of training.

This study uses LSTM model as the main model in evaluating performance on test dataset. Table 4 presents the results for the overall performance of the model, the level of aspect recognition, and the ability to classify sentiment based on the aspect. The model performance evaluation metrics include Accuracy, Precision, Recall, and F1-score.

The overall performance of the model achieved an Accuracy of 93.5% and an F1-score of 88.7%. This result demonstrates that the model is capable of accurately classifying aspects and sentiments, with a balance between Precision and Recall of 88.0% and 88.5%, respectively.

For the model's performance in aspect detection, it achieved high results with an Accuracy of 81.5% and an F1-score of 89.8%. This indicates that the model is capable of detecting aspects with high accuracy, with Precision and Recall scores of 88.1% and 91.6%, respectively.

In sentiment classification, the model achieved an Accuracy and F1-score of 90.6% each. The high Precision score of 90.8% and Recall of 90.6% indicate that the model is effective at classifying sentiment, with minimal misses or misclassifications.

Table 4. Evaluation of Overall Performance, Aspect Detection, and Sentiment Classification Metrics for the Model.

	Overall Performance	Aspect Detection	Sentiment Detection
Accuracy	93.5%	81.5%	90.6%
Precision	88.0%	88.1%	90.8%
Recall	88.5%	91.6%	90.6%
F1-score	88.7%	89.9%	90.6%

Table 5 shows the results of evaluating the performance of the model in terms of recognition on each aspect and the classification of emotions for each aspect.

Table 5. The performance evaluation metrics (%) for aspect recognition and the sentiment classification performance of the model on each aspect.

Aspect	Aspect Detection				Sentiment Detection			
	Acc	Pre	Recall	F1	Acc	Pre	Recall	F1
BATTERY	96.7	94.8	98.1	96.4	90.0	91.0	90.0	91.0
CAMERA	97.4	93.0	97.6	95.2	94.0	95.0	94.0	94.0
DESIGN	95.5	87.5	87.9	97.7	94.0	94.0	94.0	94.0
FEATURES	91.2	82.9	91.4	86.9	88.0	89.0	88.0	89.0
GENERAL	85.4	86.3	90.9	88.5	83.0	83.0	83.0	83.0
PERFORMANCE	90.1	90.0	91.4	90.7	84.0	84.0	84.0	84.0
PRICE	95.0	89.4	91.2	90.3	90.0	90.0	90.0	90.0
SCREEN	96.1	81.7	86.6	84.1	95.0	95.2	95.0	94.9
SER&ACC	90.4	81.4	82.9	82.2	89.0	89.0	89.0	89.0
STORAGE	99.3	75.0	66.6	70.5	99.0	99.0	99.0	99.0

The results in Table 5 indicate that the model performs well in aspect detection and sentiment classification. For the model's aspect recognition ability, based on Accuracy and F1-score metrics, we can see that aspects such as BATTERY, CAMERA, FEATURES, PERFORMANCE, and PRICE are well-recognized by the model, with Accuracy and F1-score metrics all exceeding 90%. Aspects with stable Accuracy or F1-score metrics around 80% include DESIGN, GENERAL, SCREEN, and SER&ACC. For the STORAGE aspect, although the Accuracy is very high at 99.3%, the F1-score is only

70.5% and Recall is 66.6%. This suggests that the model sometimes overlooks the STORAGE aspect and mainly focuses on aspects with a larger number of samples.

In terms of sentiment classification, the model performs well on aspects such as BATTERY, CAMERA, DESIGN, PRICE, SCREEN, and STORAGE, with Accuracy exceeding 90%. On the other hand, aspects like FEATURES, GENERAL, PERFOR-MANCE, and SER&ACC have both Accuracy and F1-score metrics above 80%, with FEATURES and SER&ACC results approaching 90%. Overall, the model has achieved excellent performance in sentiment classification, showing impressive results across all aspects.

5 Discusion

Overall, the models in this study has achieved good performance in aspect detection and sentiment classification on text, outperforming traditional deep learning methods such as LSTM or neural networks. Specifically, our method achieved F1-scores of 89.9% for aspect detection and 90.6% for sentiment classification, higher than the BiLSTM-based method used by Luc et al., which achieved F1-scores of 84.48% for aspect detection and 63.06% for sentiment classification. Our results also surpass the neural network method employed by Thin et al., which had F1-scores of 78.66% for aspect detection and 73.16% for sentiment classification. Compared to newer methods, our model performs similarly to the BERT-based approach used by Mickel et al., and shows superior results compared to earlier traditional methods.

However, there are two issues that need to be addressed to improve the model. First, the Accuracy score in aspect detection indicates that the model needs improvement to better identify all aspects in the text. Second, although the Accuracy for detecting the STORAGE aspect is 99.3%, the Precision, Recall, and F1-score are only around 70.0%. This may be due to class imbalance in the data, as the STORAGE class has the fewest samples, making it more challenging for the model to learn to identify this class and leading to a greater focus on classes with more samples. Our results are consistent with the method used by Luc et al., which also showed that the model struggled with detecting the STORAGE aspect.

6 Conclusion

This study proposed a method using transfer learning by combining the large language model BERT with LSTM to simultaneously perform two tasks: aspect detection and sentiment classification in a single task. This differs from traditional methods, which separate the task into two distinct steps: detecting aspects first, followed by sentiment classification.

The results of the study show that our proposed method outperforms traditional methods when applied to the same dataset. Furthermore, this method demonstrates higher stability when working with Vietnamese, which has many complex grammatical features.

However, this study has some limitations. First, our model struggles with handling sentences where there is an imbalance in the number of samples between classes. To address this, we propose using data augmentation techniques or collecting additional data from similar sources. Second, the dataset used in this study is limited in scale and diversity. While the results on this dataset are promising, the model's generalization ability on other data or from different sources may be affected. Therefore, a future research direction is to expand the dataset and test the model on various datasets and languages.

Finally, another promising avenue to explore is the use of more advanced modern language models combined with traditional methods to leverage the strengths of both approaches. We also recommend further research into optimizing model parameters to improve performance even more.

References

1. Hoang, M., Bihorac, O.A., Rouces, J.: Aspect-based sentiment analysis using BERT. In: Proceedings of the 22nd Nordic Conference on Computational Linguistics, pp. 187–196, Turku, Finland. Linköping University Electronic Press (2019)
2. Mai, L., Le, B.: Aspect-based sentiment analysis of vietnamese texts with deep learning. In: Nguyen, N., Hoang, D., Hong, TP., Pham, H., Trawiński, B. (eds.) Intelligent Information and Database Systems. ACIIDS 2018. Lecture Notes in Computer Science, vol. 10751. Springer, Cham (2018). https://doi.org/10.1007/978-3-319-75417-8_14
3. Minaee, S., et al.: Large language models: a survey. arXiv preprintarXiv:2402.06196 (2024)
4. Luc Phan, L., et al.: Sa2sl: from aspect-based sentiment analysis to social listening system for business intelligence. In: Knowledge Science, Engineering and Management: 14th International Conference, KSEM 2021, Tokyo, Japan, 14–16 August 2021, Proceedings, Part II 14, pp. 647–658. Springer (2021)
5. y Arcas, B.A.: Do large language models understand us? Daedalus **151**(2), 183–197 (2022)
6. Van Thin, D., Nguye, V.D., Van Nguyen, K., Nguyen, N.L.T.: Deep learning for aspect detection on Vietnamese reviews. In: 2018 5th NAFOSTED Conference on Information and Computer Science (NICS), pp. 104–109. IEEE (2018)
7. Van Thin, D., Le, L.S., Nguyen, H.M., Nguyen, N.L.T.: A joint multi-task architecture for document-level aspect-based sentiment analysis in Vietnamese. IJMLC **12**(4) (2022)
8. Devlin, J.: Bert: pre-training of deep bidirectional transformers for language understanding. arXiv preprintarXiv:1810.04805 (2018)
9. Hochreiter, S.: Long Short-term Memory. Neural Computation MIT-Press (1997)
10. Si, T.N., Van Hung, T.: A study on sentiment analysis combine matrix factorization for ratting product in Vietnamese. In: 2021 21st ACIS International Winter Conference on Software Engineering, Artificial Intelligence, Networking and Parallel/Distributed Computing (SNPD-Winter), pp. 57–62. IEEE (2021)
11. Chung, J., Gulcehre, C., Cho, K., Bengio, Y.: Empirical evaluation of gated recurrent neural networks on sequence modeling. arXiv preprintarXiv:1412.3555 (2014)
12. Cho, K.: Learning phrase representations using RNN encoder-decoder for statistical machine translation. arXiv preprint arXiv:1406.1078 (2014)
13. Vaswani, A., et al.: Attention is all you need. In: Advances in Neural Information Processing Systems (NIPS 2017) (2017)

Theorical Contributions for Responsible AI

Deep CNN for Remaining Useful Life Prediction: An XAI Approach Using SHAP for Model Interpretation

Le Hoang Nguyen[1,2]([⊠]), Quoc-Thông Nguyen[2], Kim Duc Tran[1,2],
Huu Du Nguyen[3], Sébastien Thomassey[1], Xianyi Zeng[1], and Kim Phuc Tran[1]

[1] Univ. Lille, ENSAIT, ULR 2461 - GEMTEX - Génie et Matériaux Textiles,
59000 Lille, France
`le-hoang.nguyen@ensait.fr`
[2] International Chair in DS & XAI, International Research Institute for Artificial
Intelligence and Data Science, Dong A University, Danang, Vietnam
[3] Faculty of Mathematics and Informatics, Hanoi University of Science and
Technology, Hanoi, Vietnam

Abstract. In a smart manufacturing setting, maintaining the stability and uninterrupted operation of machinery and equipment is a fundamental goal of the production process. Historically, equipment dependability has been preserved by monitoring usage duration and compliance with operational schedules to reduce the likelihood of unforeseen system breakdowns. Recent breakthroughs in artificial intelligence algorithms have proposed several machine learning and deep learning models as viable methods for evaluating equipment's remaining usable life. This essential method enables predictive maintenance. This research utilizes a Deep Convolutional Neural Network (DCNN) model to forecast the Remaining Useful Life (RUL) of motors in the aerospace sector. Moreover, improving the transparency and dependability of deep learning models is essential for assuring trustworthy decision-making. Consequently, Explainable Artificial Intelligence (XAI) methodologies are employed to enhance the interpretability of the DCNN model. The efficacy of the suggested strategy is shown using a reputable dataset, namely the C-MAPSS dataset by NASA.

Keywords: Remaining Useful Life · Deep Convolution Neural Network · Explainable Artificial Intelligence · Predictive Maintenance

1   Introduction

Predictive maintenance is increasingly regarded as essential to modern maintenance management strategies. By harnessing real-time data from sensors and measurement devices, alongside machine learning algorithms, it is possible to predict the optimal time for maintenance, thereby averting unexpected failures [23]. Continuous data collection from common sensors—such as temperature, vibration, and pressure—installed on equipment facilitates the monitoring

S. Thomassey et al. (Eds.): RAIDS 2024, LNICST 673, pp. 165–175, 2026.
https://doi.org/10.1007/978-3-032-14055-5_12

of potential anomalies [11]. The proliferation of IoT systems in industrial settings enables the real-time acquisition of sensor data during system operation, which is subsequently employed in machine learning models to predict potential equipment malfunctions [20]. Techniques used for RUL prediction, based on sensor data features, include linear regression models [12], and neural network models such as deep neural networks (DNN) and recurrent neural networks (RNN), which are capable of processing vast datasets and autonomously extracting relevant features for complex RUL prediction tasks. Notably, deep learning architectures like Long Short-Term Memory (LSTM) and Gated Recurrent Unit (GRU) networks excel at capturing temporal patterns and relationships within data, thereby facilitating more accurate RUL predictions [6]. In contrast to anomaly detection, RUL refers to the remaining operational time before a machine requires maintenance or replacement. By accounting for RUL, engineers can schedule maintenance, optimize operational efficiency, and avoid unscheduled downtime. RUL predictions focus on identifying events that precipitate rapid equipment degradation, enabling long-term asset management without premature replacement, which can otherwise result in unnecessary waste. In contrast, the equipment remains functional [2]. Traditional methodologies for RUL prediction, such as the application of distribution and regression models, have been widely utilized to estimate the lifespan of products [15]. Survival analysis techniques have also been applied for RUL estimation [10]. These methods often necessitate modeling failure processes, which can present challenges when applied to complex systems with multiple failure modes. The rise of smart sensors and the Internet of Things (IoT) has introduced numerous opportunities for leveraging machine learning models to perform RUL predictions based on sensor data.

The advent of advanced sensor technologies has been instrumental in facilitating the collection of real-time operational data, which can be used by machine learning algorithms to forecast potential equipment failure. A recent approach involves employing neural networks to predict RUL from engine sensor data [8], with a notable method utilizing recurrent neural networks (RNN) to forecast RUL in aircraft engines [9]. These models have demonstrated the capability to learn complex data features without relying on predefined damage distributions. Recently, deep learning models have exhibited remarkable success in RUL prediction tasks. Convolutional Neural Network (CNN) models have been employed to extract features from sensor data [22], while Long-Term Recurrent Neural Networks (LTRNN) have been used to capture temporal dependencies in sensor data sequences [21]. These models are adept at capturing non-linear and dynamic relationships between input features and RUL. Moreover, certain studies have explored integrating various methods to capitalize on their respective strengths, such as a hybrid approach combining neural networks and survival analysis models for RUL prediction [18]. These approaches facilitate the simultaneous extraction of quantitative and qualitative data information.

However, machine learning models that make predictive decisions are still a "black box" that does not yet explain the reasons for its decisions. Therefore, integrating explainable models (XAI) aims to increase transparency and promote greater practical applicability in industrial environments [7]. In this study, we will explore the performance of the DCNN algorithm in predicting RUL. In addition, we also aim to explain the model's behavior using XAI techniques, namely SHAP. The performance is tested on the CMAPSS dataset, a widely used dataset in research related to RUL prediction. We determine which sensor parameters cause motor degradation. This is considered to create reliability for users when applying artificial intelligence models to actual production.

The rest of this paper is organized as follows: in Sect. 2, we review the literature related to the use of DCNN in the field of RUL prediction and analyze the use of Explainable in explaining the decision-making reasons of machine learning models, Sect. 3 Description of the C-MAPSS dataset for preprocessing on each dataset. Section 4 Applying the Deep Convolution Neural Network algorithm to the C-MAPSS dataset to predict the Remaining Useful Life of the Engine on the datasets and showing the decision-making reasons from the dataset. Finally, some conclusions and comments are presented.

2   The Proposed Approach for Remaining Useful Life Prediction

In this section, we briefly describe the DCNN model to exploit important features from sensor data, and the methods using XAI to make the model's decisions transparent in predicting RUL.

2.1   Deep Convolution Neural Networks Model

The Deep Convolutional Neural Network (DCNN) represents a specialized deep neural network where convolutional layers extract features from input data. Convolutional Neural Networks (CNNs) are traditionally associated with image and video processing tasks, such as object recognition and image classification. CNNs can capture signal representations across various scales through a multi-layered deep architecture. In the context of time series data, CNNs have been utilized to address RUL estimation by tracking target values across sensor time series [14]. For the RUL estimation problem, the DCNN model has been adapted to handle multivariate temporal data. Rather than being limited to image classification tasks, DCNNs are applied to process data originating from multiple sensors in a regression framework to predict the remaining useful life of the equipment. The architecture of DCNN typically includes convolutional and pooling layers interspersed with activation functions such as sigmoid or ReLU, which enable the model to automatically extract significant features from temporal data [18]. In contrast to traditional models applied to each sensor in isolation, DCNNs employ bidirectional filters to simultaneously process temporal data from multiple sensors, thereby generating unified feature maps that represent the entire system. These extracted features are subsequently passed to the final regression layer, where RUL is estimated using a squared error loss function (Fig. 1).

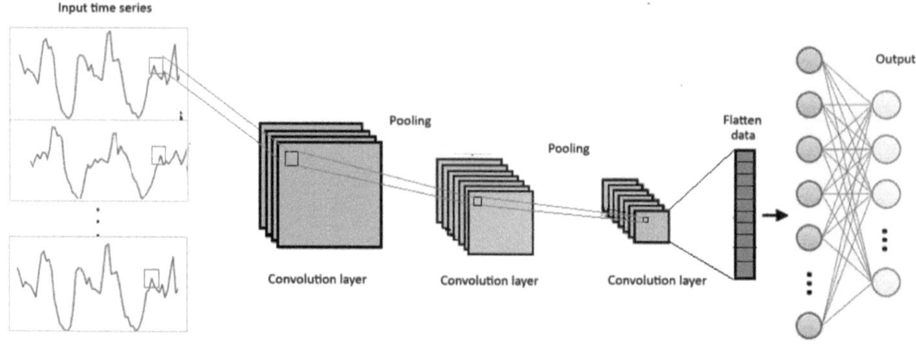

Fig. 1. Deep convolution neuron network architecture

2.2 Explainable Artificial Intelligence for Predictive Models

Machine learning models are usually considered "black boxes" as the model's decision-making process is difficult for users to understand. Recently, many studies using explainable artificial intelligence have shown many reliable benefits when using artificial intelligence in business maintenance tasks.

In predictive maintenance of remaining equipment life in plants, predictive models must explain the rationale for decisions, which is essential to give operators confidence before decisions are made, saving businesses a lot of money. Using artificial intelligence models to collect data from sensors and process and analyze that data to predict the RUL of equipment is essential in industry [5,23]. Explainable AI (XAI) is a technique that focuses on making machine learning models easier to understand for users. There are many methods used for the interpretation of AI models, we list some commonly used methods for the interpretation of AI models. XAI can be divided into two categories (Fig. 2):

- Transparent models: These models, such as Linear Regression and Decision Trees, have built-in features that explain their decisions.
  - Decision Trees: This method is easy to understand by drawing a tree diagram, showing how decisions are made at each node [3].
  - Linear Regression: Coefficients represent the influence of each input variable on the output.
- Black-box models with post-hoc explainability: These techniques develop separate models to explain the predictions of the original model.
  - Local Interpretable Model-agnostic Explanations (LIME): creates simple models near each prediction to explain the complex model's decision [17].
  - Shapley Additive exPlanations (SHAP) relies on game theory to allocate model output to each input feature, providing an approach to measure the importance of each feature [16].
  - Gradient-weighted Class Activation Mapping (Grad-CAM): Creates activation zones to see the regions of the image that the neural network focuses on when making decisions [19].

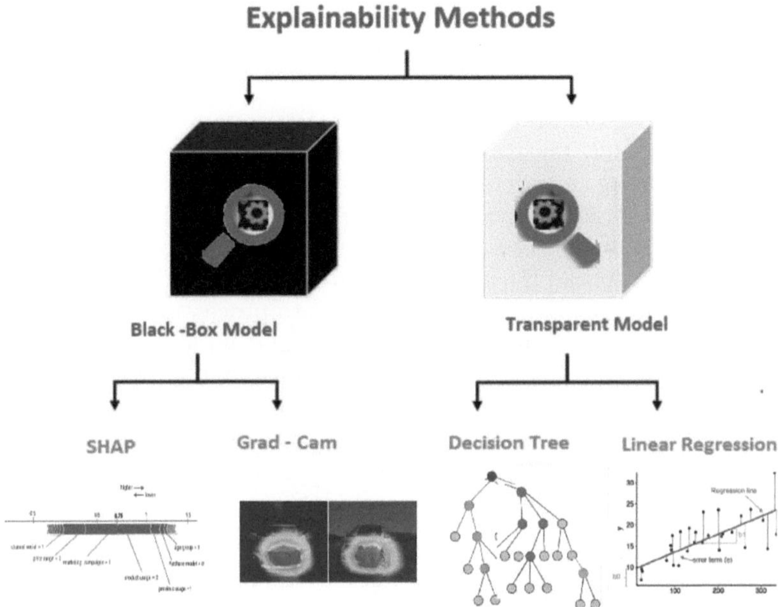

Fig. 2. Explainable Artificial Intelligence Techniques

These methods provide a range of tools to make AI models more transparent and explainable, helping users better understand how the models work and make decisions.

3   Experimental Results and Discussions

We continue on presenting the application of the DCNN model on predicting the RUL of the jet engine on the C-MAPSS simulated data, and the use of XAI to explain the model's decisions.

3.1   Data Overview

With the theoretical foundation introduced above, we will present the dataset used to deploy the RUL prediction model in this part. This dataset will be the benchmark for evaluating the accuracy and usability of the models we implement. To investigate the proposed method's performance, we use the C-MAPSS dataset developed and published by NASA (National Aeronautics and Space Administration) in this study. This dataset simulates the performance of turbofan engines under various operating conditions and with different levels of wear and damage. It provides a data foundation for developing and testing fault diagnosis and prediction algorithms and predictive maintenance in the study of RUL prediction of turbofan engines [1].

The C-MAPSS dataset is divided into four subsets: FD001, FD002, FD003, and FD004, each subset corresponding to different operating conditions and failure modes. Each dataset has a training set and a test set. This data is collected from 21 sensors to monitor the operation of the running motors from normal operation to failure [13]. The time at which the motors start to degrade until they cause system failure is considered to be the task of estimating the time RUL of the device. The actual RUL dataset is also provided in Nasa's C-MAPSS dataset. Many studies use this dataset to estimate the RUL of equipment. Each data set in the C-MAPSS set has an ID, cycle, 3 operating modes, and parameters from 21 sensors. We remove some sensor measurements to process the data before being trained on the model because they provide invalid information with constant output throughout the motor's life. This ensures that it does not affect the results but makes the training and prediction speed fast. Then we normalize the data to help the model learn more effectively. The RUL value is calculated by subtracting the current cycle number from the maximum cycle number of the motor [1, 18]. The information table of each data set is shown in Table 1.

Table 1. Information of C-MAPSS dataset

Datasets	FD001	FD002	FD003	FD004
Engine for training	100	260	100	249
Engine for testing	100	259	100	248
Operating conditions	1	6	1	6
Fault modes	1	1	2	2

After preprocessing, the data is normalized and formatted into a time series to match the structure of the DCNN model. The use of the DCNN model involves different parameters such as convolution layer, pooing, and number of neurons in each layer output and it greatly affects the accuracy of the model, the combination of values with the best results is recorded. Predictive model with DCNN architecture with multiple convolutional layers to extract features from time series data. The layers in the model will be added in order from beginning to end, the first convolutional layer has 32 the number of filters of the convolutional layer, and 8 the size of the length of the sliding window on the features. The ReLU activation function is used to create non-linearity for the model. The next layers are three consecutive convolutional layers with 64 in the second and third layers, the number of filters is increased to 64 to learn more complex features. In the fourth layer, the number of filters is reduced to 32, helping the model synthesize information after going through many convolutional layers. These layers still use the kernel size 8 and the ReLU activation function as the first layer. This pooling layer selects the largest value in each feature map. It helps reduce the output data's dimension, synthesizing the entire series's information. The dropout layer with a 50% rate reduces overfitting by randomly dropping 50 % of the neurons during training. The output fully connected layer with 100 neurons, the RMSprop optimizer with a learning rate of 1e-5, because the input data has complex nonlinear properties, the optimization algorithm is designed to adjust the learning rate automatically. The Huber loss function combines the mean squared and mean absolute error for prediction with noisy data.

3.2 The Performance of the Deep Convolution Neural Network Model

We run the DCNN model on all 4 datasets, the RMSE prediction results are shown in Table 2. The RMSE error ranges from 21 to 35.

We also present the RSME results of two other models, LSTM [4] and AGCNN [1]

In the CMAPSS datasets, FD001 and FD003 Datasets have one operating condition and one failure mode. FD002 and FD004: These datasets have multiple operating conditions and multiple failure modes, thus being more complex. In FD004, DCNN gives the lowest RMSE prediction result (21.01), showing strong prediction ability with complex data sets. DCNN is also preferred for use in problems related to time series data and RUL prediction, especially when the data is highly complex.

Table 2. Root mean square error of RUL prediction on four datasets.

Method	RMSE			
	FD001	FD002	FD003	FD004
DCNN	27.06	34.47	23.35	21.01
LSTM	16.14	24.49	16.18	28.17
AGCNN	12.42	19.43	13.39	21.50

The results of individuals with different IDs of engines are shown in Fig. 3 with the difference between the predicted and actual values in predicting RUL. The actual RUL value is the orange line, the value given before the motor degrades is 125 and remains the same for a long time, then it gradually decreases to 0 to indicate the time when the motor reaches the end of its life cycle, the predicted line fluctuates and decreases over time when the motor tends to degrade to the end of its life cycle. In the next section, we explain the decision-making reasons for the predictive model to increase reliability when using artificial intelligence. We use the DCNN model and then apply XAI to this model.

3.3 Exploring XAI Techniques for Enhancing Model Transparency and Interpretability for the Deep Convolution Neural Network Model

In this study, we apply machine learning techniques to predict RUL and use explanation AI (XAI) to analyze factors affecting RUL. As discussed above, the

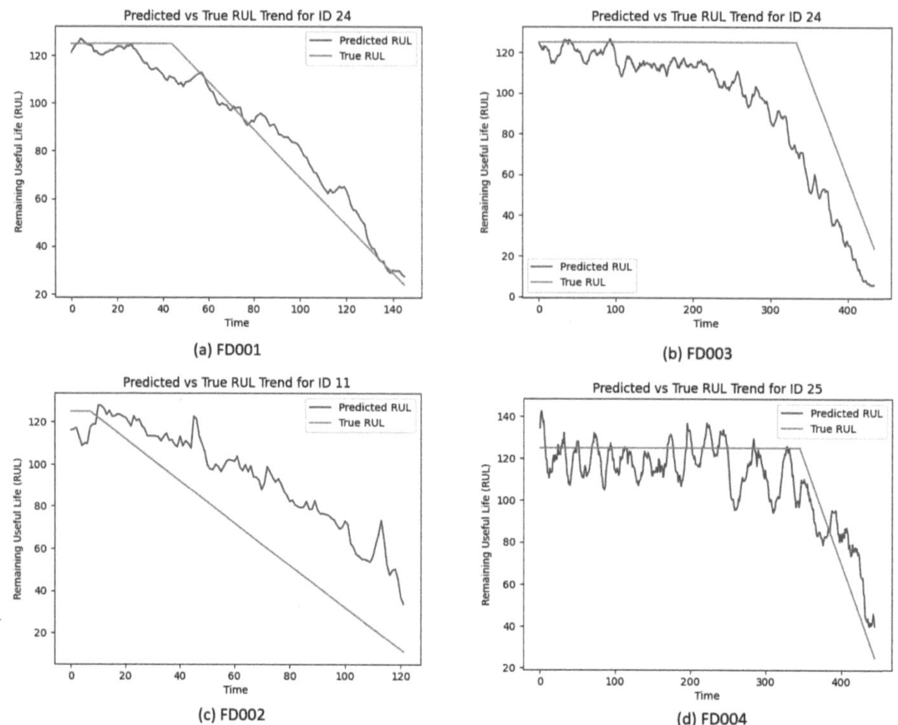

Fig. 3. Predicted - True RUL Trend for other ID of datasets

SHAP chart is a transparent method to understand better how the model makes decisions, helping to explain why the model predicts the RUL value. This chart shows how much each sensor affects the model's output value. In this section, we use the SHAP value to run on each CMAPSS dataset, which is a multivariate dataset, to examine and explain the contribution of features in the machine learning models and why the model makes predictions during execution. The distribution of individual dataset SHAP values in DCNN model. In Fig. 4, (a) is the distribution of sensors in the FD001 dataset, (b) is the distribution of sensors in the FD002 dataset, (c) is the distribution of sensors in the FD003 dataset, (d) is the distribution of sensors in FD004 dataset

The results are shown in Fig. 4, (a) for the FD001 dataset, which is explained as follows: The average predicted value is 89.28, the final predicted value of the model after all the features have been considered. Sensors like sensor12, sensor8, sensor20, etc., increased the predicted value, but sensors like sensor21 and sensor9 decreased the predicted value, creating an overall balance, making the final prediction remain the same as the base value. Sensors with longer bars have a greater influence on the prediction. This graph helps us understand better how the sensors contribute to the model's prediction, which helps explain the

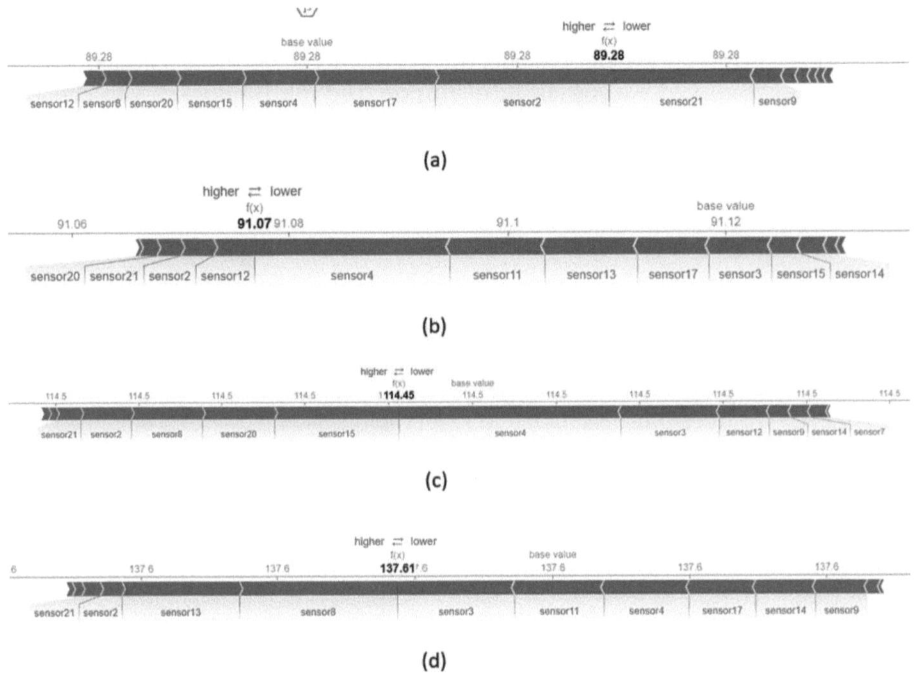

Fig. 4. The distribution of individual dataset SHAP values in DCNN model

complexity of the model's decision-making. The same results can also be seen in figures (b) for dataset FD002, figure (c) for dataset FD003, and figure (d) for FD004.

4   Conclusion

In this study, we used a DCNN model combined with model explainability techniques to predict the RUL of a turbofan engine using the C-MAPSS dataset. The results show that our model can accurately predict RUL while providing detailed information about the sensors and factors that most influence the engine degradation process.

Using SHAP to explain the model predictions allowed us to identify the critical sensors that affect the engine and point out which points of the engine affect its useful life, which helps improve maintenance efficiency and optimize engine operations. In addition, the results help enhance data-driven decision-making in industrial equipment management.

To develop and improve this research in the future, we propose the following directions: Continue to optimize the DCNN model by adjusting the hyperparameters and experiment with other network structures, such as GRU (Gated Recurrent Units) or Transformers, to compare the performance. Integrate more advanced deep learning techniques, such as hybrid models combining DCNN with other machine learning models, to improve the RUL prediction ability. Other methods in XAI can be used to support the interpretation and decision-making process based on the model results. By implementing these directions, we expect to improve the performance of the RUL prediction model and contribute to the development of advanced maintenance applications in the industry.

References

1. Asif, O., Haider, S.A., Naqvi, S.R., Zaki, J.F., Kwak, K.-S., Islam, S.R.: A deep learning model for remaining useful life prediction of aircraft turbofan engine on C-MAPSS dataset. IEEE Access **10**, 95425–95440 (2022)
2. Banjevic, D.: Remaining useful life in theory and practice. Metrika **69**(2), 337–349 (2009)
3. Mahbooba, B., Timilsina, M., Sahal, R., Serrano, M.: Explainable artificial intelligence (XAI) to enhance trust management in intrusion detection systems using decision tree model. Complexity **2021**, 1–11 (2021)
4. Boujamza, A., Elhaq, S.L.: Attention-based LSTM for remaining useful life estimation of aircraft engines. IFAC-PapersOnLine **55**(12), 450–455 (2022)
5. Nguyen, D.H., Tran, K.P., Zeng, X., Koehl, L., Castagliola, P., Bruniaux, P.: Industrial internet of things, big data, and artificial intelligence in the smart factory: a survey and perspective. In: ISSAT International Conference on Data Science in Business, Finance and Industry, pp. 72–76 (2019)
6. Ferreira, C., Gonçalves, G.: Remaining useful life prediction and challenges: a literature review on the use of machine learning methods. J. Manuf. Syst. **63**, 550–562 (2022)
7. Gama, J., Nowaczyk, S., Pashami, S., Ribeiro, R.P., Nalepa, G.J., Veloso, B.: XAI for predictive maintenance. In: Proceedings of the 29th ACM SIGKDD Conference on Knowledge Discovery and Data Mining, pp. 5798–5799 (2023)
8. Guo, L., Li, N., Jia, F., Lei, Y., Lin, J.: A recurrent neural network based health indicator for remaining useful life prediction of bearings. Neurocomputing **240**, 98–109 (2017)
9. Heimes, F.O.: Recurrent neural networks for remaining useful life estimation. In: 2008 International Conference on Prognostics and Health Management, pp. 1–6. IEEE (2008)
10. Jardine, A.K., Lin, D., Banjevic, D.: A review on machinery diagnostics and prognostics implementing condition-based maintenance. Mech. Syst. Signal Process. **20**(7), 1483–1510 (2006)
11. Javaid, M., Haleem, A., Rab, S., Singh, R.P., Suman, R.: Sensors for daily life: a review. Sens. Int. **2**, 100121 (2021)
12. Khelif, R., Chebel-Morello, B., Malinowski, S., Laajili, E., Fnaiech, F., Zerhouni, N.: Direct remaining useful life estimation based on support vector regression. IEEE Trans. Industr. Electron. **64**(3), 2276–2285 (2016)
13. Li, X., Ding, Q., Sun, J.-Q.: Remaining useful life estimation in prognostics using deep convolution neural networks. Reliab. Eng. Syst. Saf. **172**, 1–11 (2018)
14. Liu, R., Yang, B., Hauptmann, A.G.: Simultaneous bearing fault recognition and remaining useful life prediction using joint-loss convolutional neural network. IEEE Trans. Industr. Inf. **16**(1), 87–96 (2019)

15. Lu, C.J., Meeker, W.O.: Using degradation measures to estimate a time-to-failure distribution. Technometrics **35**(2), 161–174 (1993)
16. Lundberg, S., Lee, S.I.: A unified approach to interpreting model predictions. In: Advances in Neural Information Processing Systems, vol. 30. Curran Associates, Inc. (2017)
17. Ribeiro, M.T., Singh, S., Guestrin, C.: "Why should i trust you?": explaining the predictions of any classifier. In: Proceedings of the 22nd ACM SIGKDD International Conference on Knowledge Discovery and Data Mining, pp. 1135–1144 (2016)
18. Sateesh Babu, G., Zhao, P., Li, X.-L.: Deep convolutional neural network based regression approach for estimation of remaining useful life. In: Navathe, S.B., Wu, W., Shekhar, S., Du, X., Wang, X.S., Xiong, H. (eds.) DASFAA 2016. LNCS, vol. 9642, pp. 214–228. Springer, Cham (2016). https://doi.org/10.1007/978-3-319-32025-0_14
19. Selvaraju, R.R., Cogswell, M., Das, A., Vedantam, R., Parikh, D., Batra, D.: Gradcam: visual explanations from deep networks via gradient-based localization. In: Proceedings of the IEEE International Conference on Computer Vision, pp. 618–626 (2017)
20. Syafrudin, M., Alfian, G., Fitriyani, N.L., Rhee, J.: Performance analysis of IoT-based sensor, big data processing, and machine learning model for real-time monitoring system in automotive manufacturing. Sensors **18**(9), 2946 (2018)
21. Zhang, J., et al.: Attention-based convolutional and recurrent neural networks for driving behavior recognition using smartphone sensor data. IEEE Access **7**, 148031–148046 (2019)
22. Zhang, X., Zhou, X., Lin, M., Sun, J.: Shufflenet: an extremely efficient convolutional neural network for mobile devices. In: Proceedings of the IEEE Conference on Computer Vision and Pattern Recognition, pp. 6848–6856 (2018)
23. Zonta, T., et al.: Predictive maintenance in the industry 4.0: a systematic literature review. Comput. Ind. Eng. **150**, 106889 (2020)

Exploring the Impact of Different Signal Processing Techniques on Model Accuracy for Snore Detection on Edge Devices

Ngoc Thai Tran[1,2], Huu Nam Tran[1], Hong Hai Hoang[1], Quoc Minh Hieu Le[3], Hung Cuong Nguyen[3], and Anh Tuan Mai[1(✉)]

[1] VNU University of Engineering and Technology, 144 Xuan Thuy-Cau Giay, Ha Noi, Viet Nam
`tuanma@vnu.edu.vn`
[2] Hung Yen University of Technology and Education, Dan Tien-Khoai Chau, Hung Yen, Viet Nam
[3] HUS High School for Gifted Students, 182 Luong The Vinh-Thanh Xuan, Ha Noi, Vietnam

Abstract. This study investigates the impact of audio processing techniques on the performance of Convolutional Neural Networks (CNNs) for snore sound detection, a crucial component in diagnosing Obstructive Sleep Apnea (OSA). OSA affects a significant portion of the population and is associated with serious health risks. We evaluate various processing methods, including raw audio, spectrograms, mel spectrograms (MS), and Mel Frequency Cepstral Coefficients (MFCCs), to determine their effect on the accuracy and efficiency of CNN-based detection models. These models are designed for deployment on resource-constrained edge devices, enabling potential real-time applications. Results indicate that MS processing significantly enhances model accuracy, achieving up to 96% compared to 75% with raw data. Crucially, this improvement is achieved without compromising computational efficiency, making the approach suitable for wearable devices. These findings highlight the critical role of processing techniques in optimizing CNN performance for accurate and scalable snore detection, ultimately contributing to the development of more accessible OSA diagnostic tools.

Keywords: Snoring · Spectrogram · OSA · CNN · MS · MFCC

1 Introduction

Obstructive Sleep Apnea (OSA) is a prevalent sleep disorder characterized by recurrent episodes of breathing cessation during sleep, leading to serious health consequences such as cardiovascular disease and cognitive impairment [1–3]. Despite its widespread impact, a significant proportion of individuals with OSA remain undiagnosed due to limitations in screening tools and the complexity of traditional diagnostic methods like polysomnography [4, 5]. These conventional approaches are often costly, time-consuming, and inconvenient, necessitating the development of more accessible diagnostic tools [4].

© ICST Institute for Computer Sciences, Social Informatics and Telecommunications Engineering 2026
Published by Springer Nature Switzerland AG 2026. All Rights Reserved
S. Thomassey et al. (Eds.): RAIDS 2024, LNICST 673, pp. 176–187, 2026.
https://doi.org/10.1007/978-3-032-14055-5_13

Recent advances in artificial intelligence (AI) and machine learning offer promising solutions for improving the accuracy and efficiency of OSA detection. These include AI-powered analysis of medical images and the development of wearable devices that monitor physiological signals [6]. Furthermore, various machine learning algorithms, such as Convolutional Neural Networks (CNNs), have demonstrated high accuracy in OSA detection [7].

To enhance the performance of CNNs for snore-based OSA detection, effective audio signal processing is crucial. This study systematically evaluates the impact of four distinct audio processing techniques: raw audio data, spectrograms, mel spectrograms (MS), and Mel Frequency Cepstral Coefficients (MFCCs). By analyzing the influence of these methods on CNN performance, we aim to optimize snore detection accuracy and efficiency, particularly in resource-constrained environments.

2 Datasets and Methods

2.1 Datasets

The process of collecting snoring and non-snoring sound data was conducted using smartphones for recording. The snoring sounds were recorded by an individual independent of the volunteer. After recording, the audio files were segmented into 1-s files and manually labeled by the authors. This labeling process involved listening to the audio files and marking the start and end times within each 1-s segment for snoring events, as well as periods of silence or other non-snoring sounds. To ensure the reliability of the labels, a subset of the recordings was reviewed by a second technician, and any discrepancies were resolved through consensus.

The final dataset consisted of two classes: snoring and non-snoring sounds. The snoring class included various types of snoring sounds, such as light snoring, heavy snoring, and snoring with noise. The non-snoring class comprised sounds such as breathing, movement, and background noise. This dataset was then used for experiments with four processing methods as shown in Fig. 1. Our dataset consists of 400 snoring and 400 non-snoring samples collected from 20 volunteers, with each volunteer providing 20 data samples. We conducted experiments and evaluations, which indicated that collecting a large number of samples from a single volunteer does not significantly affect the results.

2.2 Methods

This study investigates the impact of audio data processing techniques on the performance of Convolutional Neural Networks (CNNs) for snore sound detection. We systematically evaluate various processing methods to identify the optimal approach for enhancing the accuracy and efficiency of snore detection using CNNs.

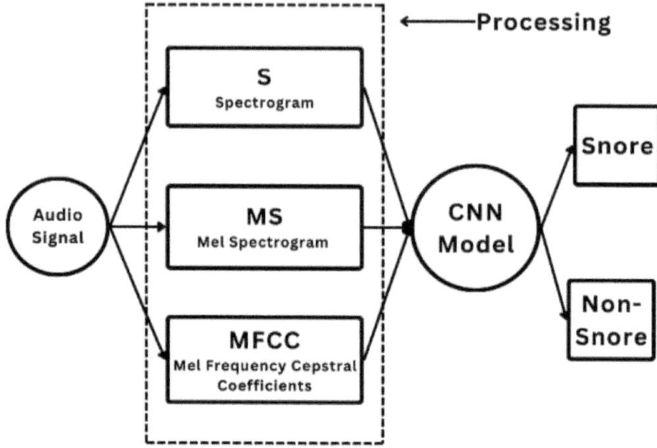

Fig. 1. Implementation diagram

As illustrated in Fig. 1, audio signal is collected by using acoustic sensor and treated by applying different data processing techniques before the data normalization and as an input of CNN model, Fig. 1.

2.2.1   Raw Data (R)

This study initially explored the use of raw audio data directly in machine learning models for snoring detection, aiming to leverage the full richness of the original signal without any preprocessing (Fig. 2).

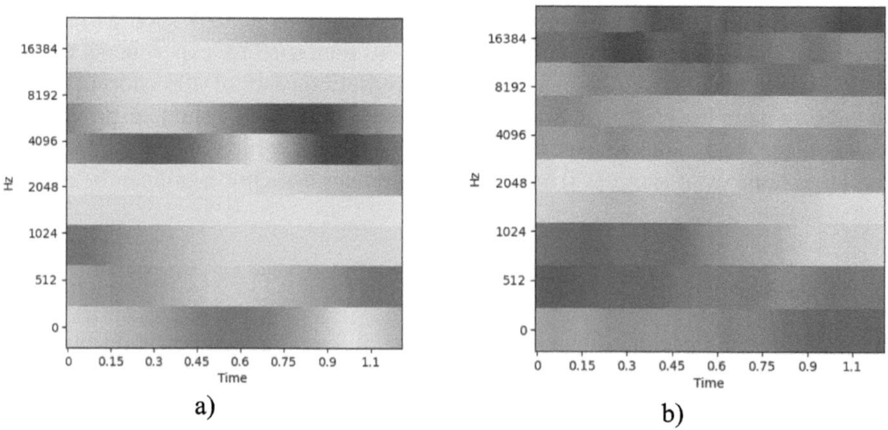

Fig. 2. Raw data of snoring a) and non-snoring b) signals

However, the high dimensionality of raw audio presented a challenge for the model input. To address this, the audio data was segmented and summed to reduce its complexity, making it compatible with a Convolutional Neural Network (CNN). Although this direct approach retains all the original information, the lack of preprocessing often limits its effectiveness compared to more sophisticated methods that refine the data before analysis.

2.2.2 Spectrogram (S)

Spectrograms provide a powerful visual tool for analyzing audio signals like snores. They reveal the hidden frequency components within a sound over time, like a musical score showing the different notes and their duration [8]. To create this visual representation, the audio signal is chopped into tiny time slices, and then a mathematical technique called a Fourier transform is applied to each slice. This process essentially decodes the sound, identifying the individual frequencies present at each moment.

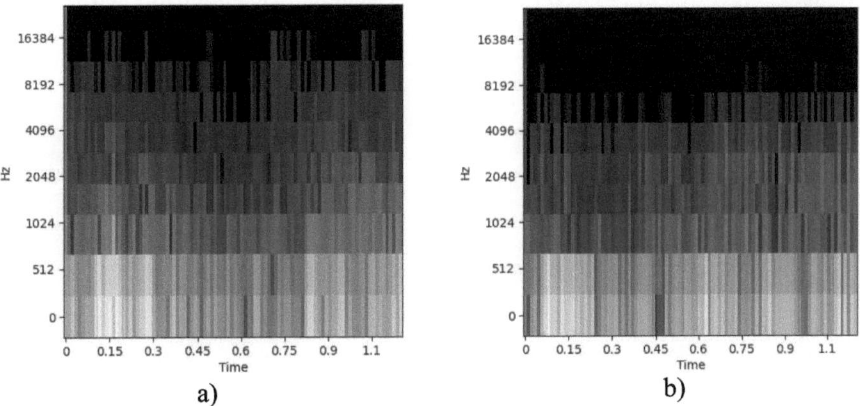

Fig. 3. Spectrogram of snoring a) and non-snoring b) signals

The result is a colorful image where time runs horizontally, frequency runs vertically, and color intensity represents the strength of each frequency at a specific time. Think of it as a heat map of sound, with brighter colors indicating stronger frequencies. This method not only preserves the temporal flow of the audio but also highlights the intricate frequency details within the snore.

Creating a spectrogram involves a few key steps. First, the audio is divided into small, overlapping segments, typically around 20–30 ms long. This overlap ensures smooth transitions and minimizes data loss. Next, each segment undergoes a Fast Fourier Transform (FFT), a rapid algorithm that converts the sound from its time-based form into its frequency-based representation. This transformation is mathematically expressed as:

$$X(k) = \sum_{n=0}^{N-1} x(n)\, e^{-j\frac{2\pi}{N}kn}$$

Next, the amplitude values, which represent the strength of the sound, are converted to decibels (dB). This conversion provides a more manageable scale for representing the wide range of sound intensities. Imagine it zooming in and out on a map; decibels allow us to see both the loud and quiet parts of the sound with greater clarity.

Finally, this processed data is assembled into a visual matrix, forming the spectrogram image. Time runs along the horizontal axis, and frequency increases along the vertical axis. The intensity or color of each point in the image reflects the loudness of a particular frequency at a specific moment. This allows us to "see" the sound evolve over time, revealing patterns and nuances that might be missed by simply listening.

Figures 3a and 3b show the spectrograms of the snoring and non-snoring signals that we transformed.

2.2.3 Mel Spectrogram (MS)

Mel spectrograms are widely employed in audio signal analysis. These spectrograms integrate the frequency spectrum with the mel scale, a non-linear scale that approximates the human perception of pitch. By providing a visual representation aligned with human auditory perception, mel spectrograms serve as effective input features for convolutional neural networks [9].

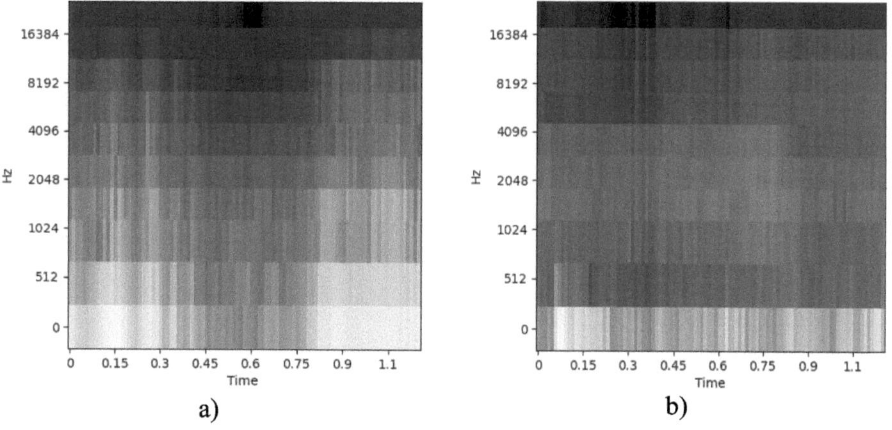

Fig. 4. Mel spectrogram of a) snoring and b) non snoring signal

To generate a mel spectrogram, the frequency axis of a traditional spectrogram is transformed using mel filter banks. This transformation maps the frequency bins onto the mel scale, ensuring that the frequency representation corresponds to the non-linear characteristics of human pitch perception. The conversion from linear frequency (f) to the mel scale (m) is mathematically defined as:

$$\text{mel}(f) = 2595\,\text{Log}_{10}\left(1 + \frac{f}{500}\right)$$

This approach enhances the representation of audio data for machine learning models by incorporating perceptually relevant features, thereby improving the accuracy and efficiency of tasks such as speech recognition and music classification.

The construction of mel spectrograms involves a crucial parameter: the number of mel bands, which typically ranges from 40 to 200. This parameter influences the resolution of the frequency representation. To generate the mel filter banks, the lowest and highest frequencies in the signal are first mapped onto the mel scale. This mel-scale range is then divided into equally spaced points, which are subsequently converted back to Hertz frequencies. Triangular filters are then constructed centered around these Hertz frequencies.

The resulting mel filter bank, represented as a matrix, is applied to the original spectrogram through matrix multiplication. This operation effectively filters the spectrogram, weighting the frequency content according to the mel scale. The resulting mel spectrogram thus provides a representation of the audio signal that approximates human auditory perception, emphasizing perceptually relevant frequency information. This makes it particularly suitable for applications involving human sound perception, such as speech recognition and music analysis.

Figures 4a and 4b show the Mel spectrograms of the snoring and non-snoring signals that we transformed.

2.2.4   Mel Frequency Cepstral Coefficients (MFCC)

Mel Frequency Cepstral Coefficients (MFCCs) are a prevalent feature representation employed in audio signal processing due to their efficacy in encapsulating the perceptually salient attributes of sound. The derivation of MFCCs involves a series of transformations applied to the short-term power spectrum of the signal. First, the power spectrum undergoes a transformation through a bank of mel filters, effectively warping the frequency axis to align with the non-linear characteristics of human pitch perception [10, 11]. This mel-scaled spectrum is then subjected to a discrete cosine transform (DCT), which serves to decorrelate the mel frequency components and produce a compact representation.

This process results in a set of MFCCs that effectively capture the spectral envelope of the sound, emphasizing the features that contribute significantly to human auditory perception. Consequently, MFCCs have proven invaluable in a wide range of audio analysis applications, including speech recognition, speaker identification, and music genre classification.

In Fig. 5, the MFCC transformation of snore and non-snore signals is shown.

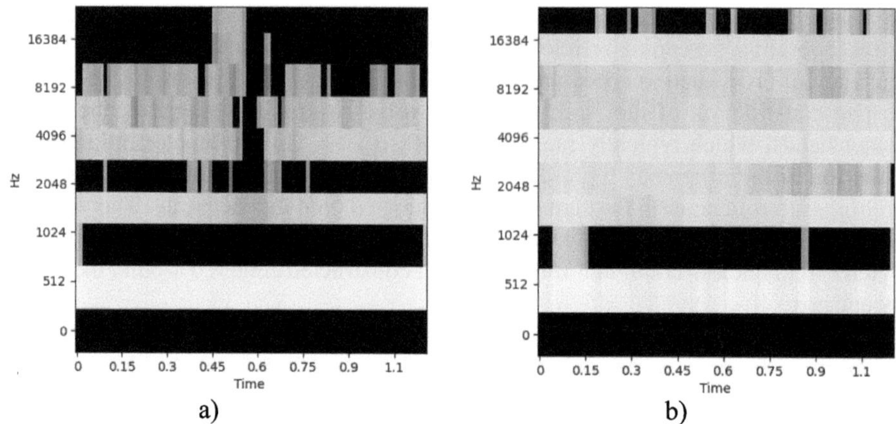

Fig. 5. Mel Frequency Cepstral Coefficients of snoring a) and non-snoring b) signals

In the realm of audio signal analysis, Mel Frequency Cepstral Coefficients (MFCCs) provide a robust representation of the spectral envelope of a sound signal, effectively capturing its perceptually relevant characteristics. This representation is achieved by first transforming the signal's power spectrum into the mel scale, a non-linear frequency scale that approximates the human auditory system's response. By emphasizing the frequency bands most salient to human perception, MFCCs have proven particularly adept at discriminating between different sound categories, including speech and non-speech signals [12, 13].

The efficacy of MFCCs stems from their ability to accentuate perceptually significant frequencies while attenuating less critical details. This attribute, coupled with recent advancements in machine learning and signal processing techniques, has led to the integration of MFCCs into sophisticated algorithms for enhanced accuracy in audio analysis [14].

The derivation of MFCCs involves a further transformation of the mel spectrogram, obtained by applying mel filter banks to the original spectrogram. This additional step entails applying the Discrete Cosine Transform (DCT) to the mel spectrogram, effectively decorrelating the mel frequency components and yielding a compact representation of the signal's short-term power spectrum. The DCT is mathematically defined as:

$$C_n = \sum_{k=0}^{K-1} M_k \cos[\frac{\pi(k + \frac{1}{2})n}{K}]$$

Applying the Discrete Cosine Transform (DCT) to the mel spectrogram yields a set of Mel Frequency Cepstral Coefficients (MFCCs) that efficiently capture salient audio features while reducing dimensionality. This compact representation is highly effective for machine learning models in applications like speech recognition and audio classification. The DCT decorrelates the mel frequency components, facilitating dimensionality reduction and enhancing computational efficiency. MFCCs retain crucial perceptual information, making them powerful discriminative features for diverse audio analysis tasks.

2.2.5 Model Built

Convolutional Neural Networks (CNNs) excel in sound signal classification due to their ability to extract discriminative features from frequency domain data. For edge computing, a streamlined CNN architecture was developed, minimizing computational demands through layer reduction and fine-tuning. Consecutive Conv2D layers enhance feature learning, while interspersed MaxPooling layers reduce dimensionality and improve noise robustness. The resulting model (Table 1) comprises 40,706 trainable parameters.

Table 1. CNN model architecture

	Type	kernel	Kernel size	Notes	Input shape
1	Conv2D	128	2×6	Activation = relu	$10 \times 100 \times 1$
2	Conv2D	16	2×2	Activation = relu	$9 \times 95 \times 128$
3	Dropout			Rate = 0.2	$8 \times 94 \times 16$
4	MaxPooling2D		2×4		$8 \times 94 \times 16$
5	Conv2D	16	2×6	Activation = relu	$4 \times 23 \times 16$
6	Flatten				$3 \times 18 \times 16$
7	Dense	32		Activation = relu	864
8	Drop			Rate = 0.2	32
9	Dense	2		Activation = softmax	32

3 Results and Discussion

The selected CNN model (Table 1) has a memory footprint of 159.01 KB, making it suitable for deployment on the chosen hardware platform: an ARM Cortex-M4 microcontroller with 256 KB of program memory. This efficient model size enables on-device implementation, facilitating real-time snoring detection within the resource constraints of the embedded system.

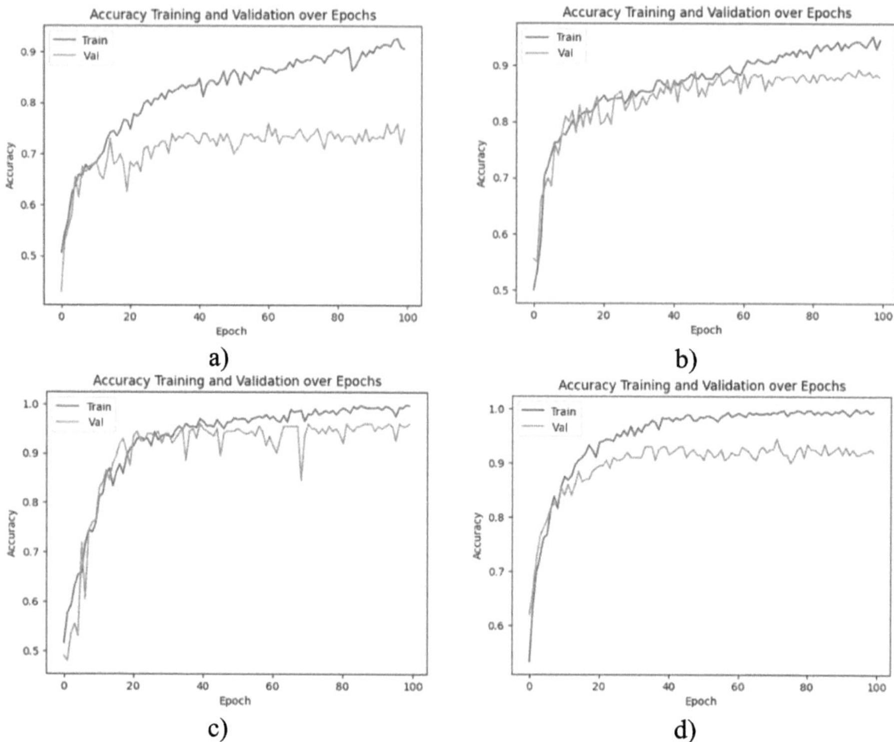

Fig. 6. Accuracy training and validation over Epochs: a-Raw data, b-Spectrogram, c-Mel Spectrogram, d-MFCC.

Analysis of the proposed methods (Fig. 6, Table 2) reveals that the Mel Spectrogram (MS) processing technique achieved the highest classification accuracy, with 96% for the validation set and 94.99% for the test set. This demonstrates the effectiveness of combining CNNs with MS for snore sound classification.

Table 2. Accuracy training, validation and test Model

Features	Train	Validation	Test
Raw data	0.9146	0.75	0.709
Spectrogram	0.9508	0.8800	0.885
Mel Spectrogram	0.9975	0.9600	0.95
MFCC	0.9967	0.9200	0.935

Evaluation of the CNN model with different processing methods (Fig. 6, Table 2) indicates high accuracy for both Mel Spectrogram (MS) and Mel Frequency Cepstral Coefficients (MFCC). While MFCCs are advantageous for applications like speech recognition due to their data compression capabilities, this compression may lead to the loss of subtle audio details crucial for snore detection. Conversely, MS preserves full time-frequency information, enabling a more comprehensive analysis of sound energy variations and facilitating the detection of subtle features critical for accurate snore identification. This advantage likely contributes to the superior performance observed with MS processing.

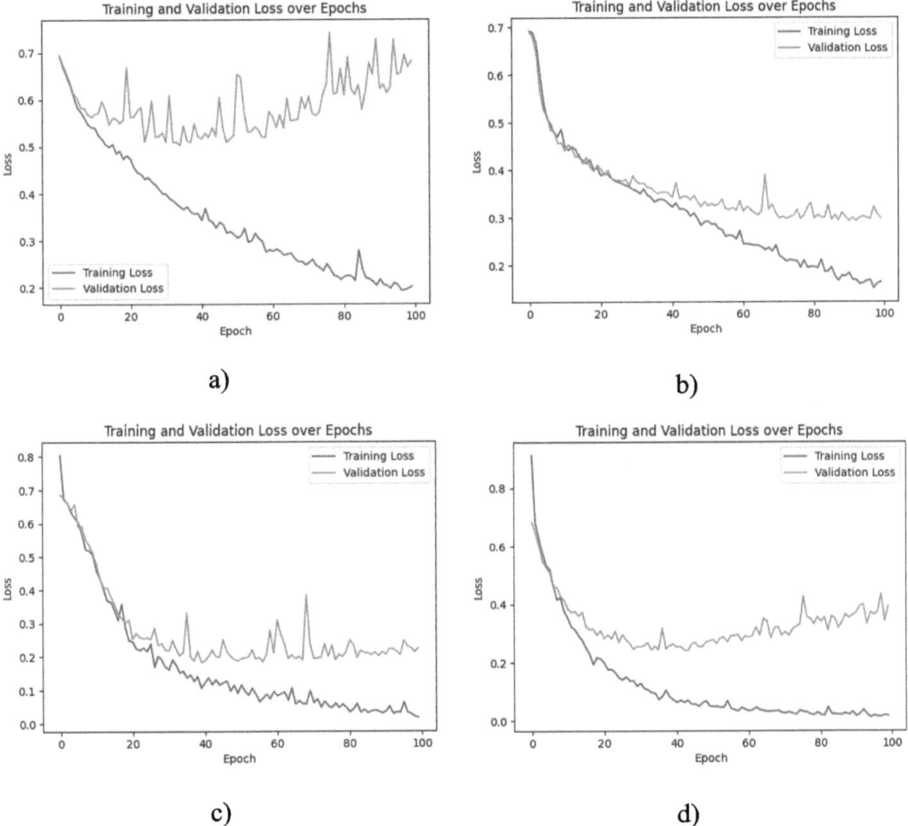

Fig. 7. Training and validation loss over Epochs. a-Raw data, b-Spectrogram, c-Mel Spectrogram, d-MFCC

Further analysis considered the model's loss rate during training, Fig. 7. The loss rates for Raw data, Spectrogram (S), Mel Spectrogram (MS), and MFCC processing were 0.693, 0.352, 0.133, and 0.277, respectively. This highlights the superior performance of MS, achieving the lowest loss rate among the evaluated methods.

4 Conclusion

This study rigorously evaluated various signal processing methods for snore detection using a consistent CNN model. Results demonstrate the superior performance of mel spectrograms, which, in conjunction with CNNs, yield high feature extraction efficacy and significantly improved accuracy compared to MFCCs, spectrograms, and raw data. The ability of mel spectrograms to capture critical low-frequency characteristics enhances the resolution and sensitivity of the CNN model for snore sound analysis. This approach, with its high accuracy and suitability for real-time implementation, represents a significant advancement towards developing edge devices for snore severity assessment and associated health risk evaluation.

Author Contributions:.   Ngoc-Thai Tran conceived of the presented idea. Huu Nam Tran and Hong Hai Hoang perfomed the the computations and verified the analytical methods. Quoc Minh Hieu Le, Hung Cuong Nguyen developed the theory and performed the computations. Anh Tuan Mai supervised the findings and was coresponding of this work. All authors discussed the results and contributed to the final manuscript.

References

1. Adams, N., Strauss, M., Schluchter, M., et al.: Relation of measures of sleep-disordered breathing to neuropsychological functioning. Am. J. Respir. Crit. Care Med. **163**, 1626–1631 (2001)
2. Reichmuth, K.J., Austin, D., Skatrud, J.B., et al.: Association of sleep apnea and type II diabetes: a population-based study. Am. J. Respir. Crit. Care Med. **172**, 1590–1595 (2005)
3. Burwell, C.S., Robin, E.D., Whaley, R.D., et al.: Extreme obesity associated with alveolar hypoventilation: a Pickwickian syndrome. Am. J. Med. **21**, 811–818 (1956)
4. Tran, N.T., Tran, H.N., Mai, A.T.: A wearable device for at-home obstructive sleep apnea assessment: State-of-the-art and research challenges. Front. Neurol. **14**, Article 1123227 (2023)
5. Ye, P., et al.: Diagnosis of obstructive sleep apnea in children based on the XGBoost algorithm using nocturnal heart rate and blood oxygen feature. Am. J. Otolaryngol. **44**(2), 103714 (2023)
6. Shajari, S., Kuruvinashetti, K., Komeili, A., Sundararaj, U.: The emergence of AI-based wearable sensors for digital health technology: a review. Sensors **23**(23), 9498 (2023)
7. Dey, D., Chaudhuri, S., Munshi, S.: Obstructive sleep apnoea detection using convolutional neural network based deep learning framework. Biomed. Eng. Lett. **8**(1), 95–100 (2017)
8. Handy, R.: Spectrograms: turning signals into pictures. J. Eng. Technol. **24**, 32–35 (2007)
9. Wyse, L.: Audio Spectrogram Representations for Processing with Convolutional Neural Networks (2017). https://doi.org/10.48550/arXiv.1706.09559
10. Abdul, Z.K., Al-Talabani, A.K.: Mel frequency cepstral coefficient and its applications: a review. IEEE Access **10**, 122136–122158 (2022)
11. Majeed, S., Husain, H., Samad, S., Idbeaa, T.: Mel frequency cepstral coefficients (MFCC) feature extraction enhancement in the application of speech recognition: a comparison study. J. Theor. Appl. Inf. Technol. **79**, 38–56 (2015)
12. Martin, A., Charlet, D., Mauuary, L.: Robust speech/non-speech detection using LDA applied to MFCC. In: 2001 IEEE International Conference on Acoustics, Speech, and Signal Processing. Proceedings (Cat. No. 01CH37221), Salt Lake City, UT, USA, vol. 1, pp. 237–240 (2001)

13. Krishna Kishore, K.V., Krishna Satish, P.: Emotion recognition in speech using MFCC and wavelet features. In: 2013 3rd IEEE International Advance Computing Conference (IACC), Ghaziabad, India, pp. 842–847 (2013)
14. Vimal, B., Surya, M., Darshan, Sridhar, V.S., Ashok, A.: MFCC based audio classification using machine learning. In: 2021 12th International Conference on Computing Communication and Networking Technologies (ICCCNT), Kharagpur, India, pp. 1–4 (2021)

The Evolution of AI Chips: From Cloud to Edge, Powering the Future of Artificial Intelligence

Nguyen Van Long, Tran Ngoc Vinh, Hoang Van Quyen,
Nguyen Thi My Duyen, Phi Van Hoa, and Mai Anh Tuan$^{(\boxtimes)}$

VNU University of Engineering and Technology, 144 Xuan Thuy-Cau Giay,
Ha Noi, Vietnam
`tuanma@vnu.edu.vn`

Abstract. Artificial Intelligence (AI) has made remarkable advancements since its inception, with AI chips playing a pivotal role in this evolution. The advancement of specialized hardware, including Graphics Processing Units (GPUs) and Tensor Processing Units (TPUs), has progressed rapidly in response to the increasing computational requirements of AI applications. Cloud-based AI leverages the vast computational resources of centralized data centers to handle large-scale data processing and complex model training, making it ideal for applications like large-scale data analytics, natural language processing, and scientific research. Conversely, Edge AI enabling real-time data processing and decision-making essential for applications with stringent latency, bandwidth, and privacy requirements, such as autonomous vehicles, health applications, and smart-city infrastructure. By examining the architectural advancements and real-world applications of both cloud-based AI and edge AI. This paper highlights the complementary strengths of these technologies and emphasizes the continued evolution and importance of AI hardware in driving innovation across various sectors, advocating for a balanced approach to enhance the full potential of both Cloud-based and Edge AI to create intelligent, responsive, and scalable solutions.

Keywords: Artificial Intelligence · Cloud-based AI · Edge AI · Healthcare

1 Introduction

Along with the innovation of the 4.0 industrial revolution, artificial intelligence (AI) is considered one of the most breakthrough technologies, driving successful digital transformation and attracting the attention and investment of tech giants.

The AI market is growing very quickly. The report estimates that its value will increase from about \$214.6 billion in 2024 to \$1,339.1 billion by 2030, with

---

VNU University of Engineering and Technology.

S. Thomassey et al. (Eds.): RAIDS 2024, LNICST 673, pp. 188–217, 2026.
https://doi.org/10.1007/978-3-032-14055-5_14

a strong annual growth rate of 35.7% during this time [1]. In comparison, Grand View Research gives an even higher prediction, saying that the global AI market could reach around \$1,811.75 billion by 2030, with an annual growth rate of 36.6% from 2024 to 2030 [2]. These forecasts point to a robust growth trend as well as a strong interest in AI technology in the near future.

The evolution of AI solutions holds promise in addressing diverse applications such as healthcare, finance, and manufacturing, offering the potential for more efficient outcomes. According to a recent McKinsey Global Survey on AI, 65% of respondents stated that their organizations have already implemented AI to gain a competitive advantage in the blockchain sector—almost twice the proportion reported just ten months earlier [3]. Furthermore, 83% of companies identified the integration of AI into their business strategies as a top priority, aiming to improve manufacturing efficiency and support long-term digital transformation initiatives [4]. For instance, Caterpillar's marine division demonstrates a substantial annual savings of \$400,000 per ship through the application of Machine Learning techniques to optimize hull cleaning frequency [5]. Furthermore, Unilever's integration of AI via Pemetrics and Higher View led to notable enhancements in their hiring process. Using AI algorithms, Unilever narrowed down a candidate pool of 250,000 to 3,500 and enabled candidates to engage with real recruiters before the final selection of 800 hires. Additionally, Unilever cut over 50,000 h in interview time over 18 months, saving 1 million pounds annually. They also achieved a remarkable 96% candidate completion rate, up from 50% using traditional methods. AI adoption resulted in a 90% reduction in time-to-hire and increased new hire diversity by 16% [6].

A remarkable illustration of AI technology is the incredible emergence of ChatGPT, which was developed by OpenAI in 2020 and has revolutionized both real-life applications and the AI market. Using the advanced Transformer architecture, specifically GPT-4, ChatGPT is outstanding in natural language processing with self-attention mechanisms for contextual understanding and human-like responses. This breakthrough technology significantly impacts various sectors, especially customer service, education, and healthcare, by providing efficient, reliable, and personalized interactions. Initially, GPT-3 set a new standard with 175 billion parameters, and GPT-4 further enhanced capabilities with a context window of up to 8,192 tokens. Its core algorithm involves pre-training on diverse text data and fine-tuning with human feedback, ensuring high quality, adaptive responses. It took ChatGPT just 5 days to reach 1 million users and 2 months to reach the milestone of 100 million users, making it the fastest-growing consumer application in history, according to a UBS study [7]. As of July 2024, ChatGPT has approximately 180.5 million users worldwide [8]. This impressive statistic highlights the widespread adoption and influence of the AI chatbot and is evidence of GPT's utility and popularity across various sectors. The success of ChatGPT demonstrates the profound capabilities and potential of modern AI systems, firmly establishing its position as a transformative tool in contemporary society (Fig 1).

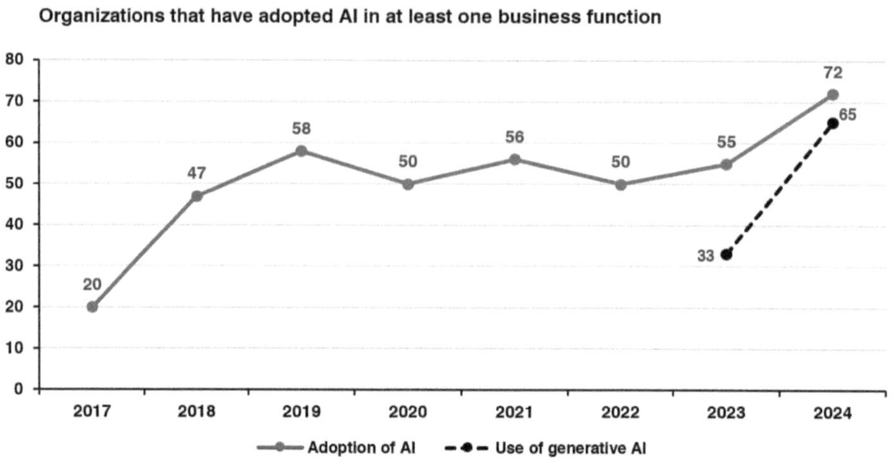

In 2017, the definition for AI adoption was using AI in a core part of the organization's business or at scale. In 2018 and 2019, the definition was embedding at least one AI capability in business processes or products. Since 2020, the definition has been that the organization has adopted AI in at least one function.

Fig. 1. Organizations Using AI in at Least One Operational Area

In practical applications, the deployment of artificial intelligence systems necessitates significant advancements in hardware to meet the increasing demands for speed and large-scale data storage, enabling significant advancements in fields such as autonomous driving, medical diagnostics,... Basically, the AI chip market can be categorized into two main segments according to their deployment methods: Cloud-based AI chips and Edge AI chips (Fig 2).

Cloud-based AI refers to artificial intelligence services and resources that are hosted on public cloud platforms. These services are accessible to users over the internet, and the AI models and data are stored on the cloud provider's servers. With advancements in hardware technology, cloud-based AI can handle extensive computational workloads, enabling the processing of large-scale data and complex algorithms [10]. The primary strengths of cloud-based AI include its scalability, substantial computational power, and centralized data management, which facilitate tasks such as deep learning model training, big data analytics, and global data accessibility [11].

Cloud-based AI refers to artificial intelligence services and resources that are hosted on public cloud platforms. These services are accessible to users over the internet, and the AI models and data are stored on the cloud provider's servers. Top cloud providers are listed in Table 1.

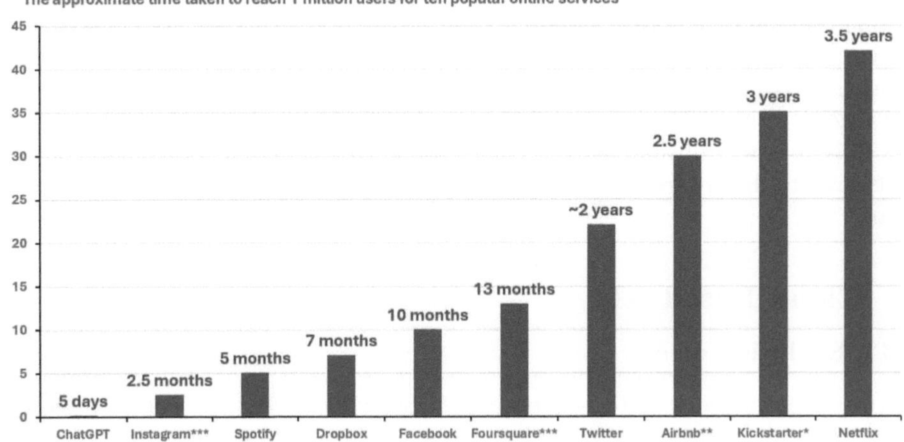

*1 million backers ** 1 million nights booked *** 1 million downloads*

Fig. 2. Approximate time to reach 1 million users for online services [9]

Table 1. Top Cloud AI service providers

No	Name	Service's link
1	AWS AI Services	https://aws.amazon.com/vi/ai/services/
2	Microsoft Azure AI	https://azure.microsoft.com/en-us/solutions/ai
3	Google Cloud AI	https://cloud.google.com/products/ai?hl=vi
4	H2O.ai	https://h2o.ai/
5	IBM Watson Studio	https://www.ibm.com/products/watson-studio
6	Oracle Cloud Infrastructure AI Services	https://www.oracle.com/artificial-intelligence/ai-services/
7	Wipro Holmes	https://www.wipro.com/ai/
8	Alibaba Cloud	https://www.alibabacloud.com/vi?_p_lc=16
9	Salesforce	https://www.salesforce.com/uk/
10	DataRobot	https://www.datarobot.com/

With advancements in hardware technology, cloud-based AI can handle extensive computational workloads, enabling the processing of large-scale data and complex algorithms [10]. The primary strengths of cloud-based AI include its scalability, substantial computational power, and centralized data management, which facilitate tasks such as deep learning model training, big data analytics, and global data accessibility [11]. As a result, cloud-based AI is widely applied in sectors requiring heavy data processing, like e-commerce, banking, and healthcare, as it enables machine learning algorithms to uncover concealed patterns and insights within large data repositories... However, the main drawbacks of cloud-based AI are higher latency and dependency on stable internet connectivity, which can impede real-time applications and data privacy [12] [13]. To address these limitations, edge AI has been developed, which processes data locally on edge devices. This approach offers low latency, enhanced data privacy, and operational reliability without relying on centralized data centers. As a result, edge AI is particularly suited for real-time applications such as autonomous vehicles, industrial automation, and real-time health monitoring, providing immediate data processing and decision-making capabilities [14].

This review will analyze the overview of AI and the development of both software and hardware in AI chip models to adapt to the needs of real applications.

2   Cloud Based AI

2.1   Evolution of AI

Artificial intelligence (AI) began to develop when scientists asked a question about designing machines with the ability to think like humans. In 1950, Alan Turing published a paper "Computing Machinery and Intelligence," in which he posed "Can machines think?" and proposed a game to evaluate what an intelligent machine is, which famously known as "The Turing test" [15]. Turing evaluated an intelligent machine as a machine that can communicate naturally, like a human. Artificial intelligence officially become a field of study in 1956 at the Dartmouth conference and millions of dollars were spent to develop a machine with human intelligence [16]. In the early days, AI researchers worked on problem-solving using symbolic methods, which helped develop early programs like the General Problem Solver [17], Natural language processing program ELIZA [18], and the discovery of the Perceptron [19], which formed the practicality of machine thinking. Despite initial successes, these early AI systems were limited in their ability to solve problems, rather than being widely applied in many areas as originally expected. interest in AI and research funding declined dramatically in the 1970s, a period known as "The AI winter" [20]. The industry turned down on artificial intelligence, but funding and research on the topic grew under different names. In the 1980s, Expert Systems flourished and revived the field of AI research. Expert Systems use knowledge from experts and follow logical steps to give answers or solve problems in specific domains [21]. MYCIN was developed in 1972 to identify infectious blood diseases and recommend the appropriate medication for them. Digital Equipment Corporation hired CMU to complete an expert system known as XCON in 1980. By 1986, it was saving the company forty million dollars a year, it was an enormous hit [22]. A lot of money was invested as a result of the AI boom, especially in Japan, where the government invested heavily in the development of the Fifth Generation Computer System [23] Its goal was to create a machine capable of human-like speech, language translation, image recognition, and argumentation. In the 21st century, the field of AI was booming thanks to the improvement in computing hardware, large data resource accessibility, and smarter algorithms. Machine Learning (ML), particularly supervised learning and neural networks, gained prominence. In 1986 Hinton, Rumelhart, and Williams published "Learning Representations by Back-Propagating Errors", allowing much deeper neural networks to be trained [24]. In 1997, IBM's Deep Blue defeated world chess champion Garry Kasparov, showcasing the potential of AI in strategic games [25]. This victory underscored the capabilities of AI in solving complex problems. A significant breakthrough came in 2012 with Geoffrey Hinton's work on deep learning, where his team achieved a substantial improvement in image recognition using deep neural networks [26]. This milestone sparked a wave of innovation, leading to advancements in various domains, including natural language processing (NLP), computer vision, and autonomous systems.

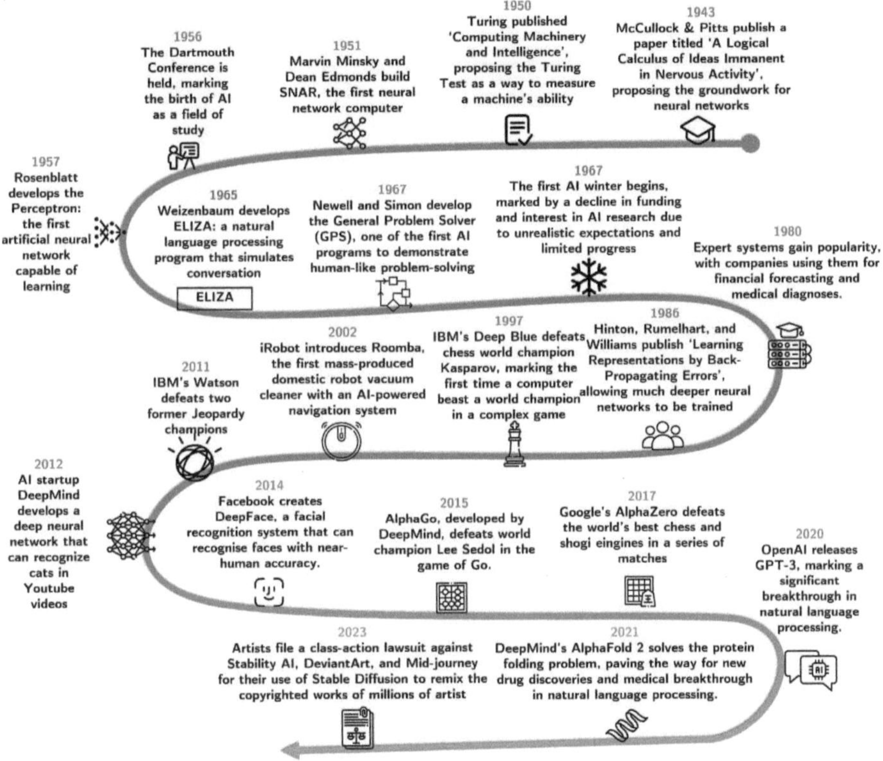

Fig. 3. Brief History of AI

Techniques like Convolutional Neural Networks (CNNs), Recurrent Neural Networks (RNNs), and Deep Neural Networks (DNNs) have achieved remarkable accuracy in these areas [27]. The transformer architecture is a deep learning architecture developed by researchers at Google based on the multi-head attention mechanism, [28] debuted in 2017 and has been covering a wide range of applications, from simple software on a smartphone to security, surveillance [29], autonomous vehicle controls [30], healthcare, robotics, web searches, computer vision, etc. Typical AI models that can be named are: ChatGPT a conversational AI model designed for interactive and natural language conversations;

Google Imagen for creating high-quality images from textual descriptions [31]; and AlphaFold for predicting protein structures with high accuracy [32]. Figure 3 shows a brief history of AI, going from a simple problem solver to a cornerstone technology. AI models are becoming more complex and data-intensive. Billions of matrix multiply-and-accumulate (MMAC) operations require a powerful hardware platform to be operated. This need for strong and efficient processing capabilities has led to significant developments in microprocessor design.

2.2 Microprocessors Development

Microprocessor improvements made the implementation of complex AI algorithms possible. The first commercial microprocessor, the Intel 4004, was introduced in 1971. With 2300 transistors, the Intel 4004 [33] was a 4-bit CPU that could process 92600 instructions per second (IPS) at a 740 kHz clock frequency. The Intel 8008, an 8-bit microprocessor, was introduced in 1972. Later, it was upgraded to the Intel 8080 in 1974, the Intel 8085 in 1976, and the Intel 8086 in 1978, which was improved to 16-bit architecture [34]. The 8086 is split into 2 functional parts: the Bus Interface Unit (BIU) and the Execution Unit (EU). This division laid the foundation for pipeline techniques. The Intel 8086 is designed to be backward compatible with the earlier Intel 8080 and 8085. The 8086's architecture set the stage for subsequent generations of Intel processors. Each new generation builds upon the 8086's architecture, also known as the x86 architecture of which powered the first PC in the world made by IBM. As the IBM PC exploded in popularity because it was a versatile, well-built computer [35], the software developers wanted to perform their programs with the IBM hardware compatibles. Intel quickly became a very powerful name in the microcomputer CPU space, and they ended up licensing out the x86 architecture to other companies such as AMD and IBM. Around that era, the Motorola 68000 laid the groundwork for full 32-bit architectures by introducing instruction fetching in multiple 16-bit word segments. The Motorola 68010 introduced the concept of virtual memory. The Motorola 68020, released in 1984, marked a milestone as the first fully 32-bit microprocessor equipped with an actual instruction pipeline and a 256-byte on-chip instruction cache. These early CPUs were designed for a wide range of applications, yet they were not powerful enough to bring AI models into practical implementation.

Researchers aimed to further enhance processing bandwidth but encountered major difficulties associated with decoding complex instruction sets. To overcome this challenge, the Reduced Instruction Set Computer (RISC) [36] has been introduced by Advanced RISC Machines (ARM). The RISC architecture and higher transistor density drove the evolution of microprocessor design with more on-chip units, deeper pipelined stages, multilevel caching, higher operating frequencies, and increased bandwidth [37]. Taking architecture optimization a step further, Stanford University introduced the "millions of instructions per second" (MIPS) architecture, which removed interlocked pipeline stages [38]. To further increase processing performance, researchers tried to increase the clock frequency because a higher clock rate equals more instructions executed. In the mid-2000s, the power dissipation barriers prevented the clock frequency from increasing any further. To achieve higher frequencies, big companies keep on shrinking feature size to the point where their chips release an enormous amount of heat, which requires impractical cooling systems. As shown in Fig. 4 the escalating power density of Intel chips was projected to result in extreme levels of heat dissipation. Remarkably, if the chip densities keep increasing, they could have reached the temperature comparable to the sun's surface [39]. To address the performance bottleneck in microprocessors, parallel computing techniques were introduced by integrating multiple closely coupled processor cores on a single chip. This multi-core CPU design allowed execution of multiple threads, significantly enhancing performance for parallelizable tasks. Multi-processor chip (MPC) architecture was further improved to quad-core and higher-core designs, initiating a new phase of performance gains with advantages in cost-effectiveness, power consumption, and applicability across diverse domains [40]. This development was crucial for AI, where tasks such as matrix multiplications and vector operations benefit immensely from parallel processing.

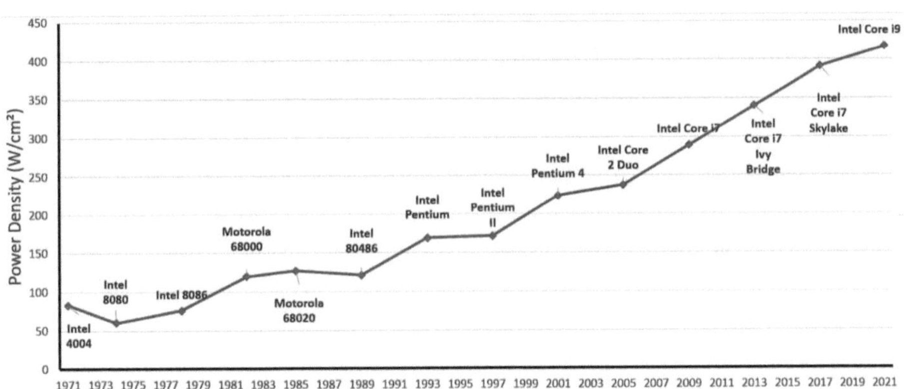

Fig. 4. Microprocessor Power density Trend

The exponential growth in transistor density, as predicted by Moore's Law [46], enabled the transition from 4-bit microprocessors to 64-bit ones, along with advanced features like pipelining, caching, virtual memory, and application-specific Very Large Instruction Word (VLIW) architectures. Table 2 summarizes some of the most prominent processors over the years. Despite their capabilities, general-purpose systems with advanced microprocessors remained unable to meet the requirements of ML and DNN workloads.

Table 2. Central Processing Unit Evolution over Time

Microprocessor	Year	Clock Speed (GHz)	Transistor Size (nm)	Caches (MB)	Cores
Intel 8086 [34]	1978	0.005	3,000	0.5	1
AMD Athlon 64 X2	2005	2	90	0.5	2
Intel Core 2 Duo	2006	2.66	65	4	2
Intel Core 2 Quad Q6600 [41]	2007	2.4	65	8	4
Intel Core i7-7700 [42]	2012	4.2	14	8	4
AMD Ryzen 7 7600x [43]	2017	5.3	5	38	6
AMD Ryzen 9 9950x [44]	2024	5.7	4	64	16
Intel Core i9 14900KS [45]	2024	6.2	10	36	24

2.3   Graphics Processing Unit

Unlike general-purpose CPUs that execute multiple tasks sequentially, AI models are inherently designed for highly parallel computation [47]. For large training datasets, AI necessitates massive parallel operations to effectively train the numerous layers of neuron networks. Certain models may comprise over 100 layers in architecture, rendering the tasks of training them unfeasible with only CPU. As a result, Graphic Processing Units (GPUs) [48], which provide extremely high levels of parallelism, have become a valuable solution for AI processing. The GPUs were designed for highly parallel applications with many computational cores and high memory access speeds [49].

Originally, GPUs were designed for 3D graphic tasks, NVIDIA's GeForce 256 is the first GPU specialized for real-time graphic processing [50]. However, GPU's architecture was the massive parallel processor for general-purpose applications from 2006, when GPU started to apply general-purpose cores [51]. Additionally, NVIDIA has developed a framework software CUDA (Compute Unified Device Architecture) [52] which was used to accelerate parallel computing. It was a specific programming language for programs designed to run on GPU with CUDA support and compatible with nearly any operating system. With the usage of the CuDNN library [53], applications such as Pytorch [54], Caffe [55], and Tensor-Flow [56] might be implemented with such complicated neural networks (NNs) thanks to NVIDIA's GPU. The CuDNN library is included in CUDA-X AI [57], a collection of libraries designed to speed up deep learning and machine learning tasks on NVIDIA GPUs.

GPUs are the current workhorses for MLs inference and especially training. They contain up to thousands of cores to work efficiently on highly parallel algorithms. In certain parts of its platform, NVIDIA has integrated Tensor cores [58] with conventional CUDA cores at the hardware level. The NVIDIA V100 architecture has 5,120 FP32 cores and 2,560 FP64 cores along with 640 tensor cores and 21.1 billion transistors. The Ampere 100 (A100) is a GPU released in 2020 by NVIDIA that targets the High Performance Computing (HPC), AI, and data center market [62]. The NVIDIA A100 GPU enhanced performance by increasing the number of cores (refer to Table 3) and offering support for

Table 3. NVIDIA's GPU Comparison

	V100 [58]	A100 [59]	H100 (PCIe) [60]	H100 (SXM5) [60]	B200 [61]
Architecture	Volta	Ampere	Hopper	Hopper	Blackwell
Launch	2017	2020	2022	2022	2024
CUDA Cores	5,120	6,912	14,592	16,896	–
Tensor Cores	640	432	456	528	–
Boost Clock (GHz)	1.53	1.41	1.62/1.755	1.83/1.98	–
INT8 TOPs	N/A	624	3,341	3,958	4,500
FP16 TFLOPs	125	312	756	989.4	2,250
TF32 TFLOPs	N/A	156	378	494.7	1,100
FP64 TFLOPs	N/A	19.5	51.2	66.9	40
FP16 TFLOPs (non-Tensor)	31.4	78	102.4	133.8	–
FP32 TFLOPs (non-Tensor)	15.7	19.5	51.2	66.9	–
FP64 TFLOPs (non-Tensor)	7.8	9.7	25.6	33.5	–
Memory	16 GB	40 GB	80 GB	80 GB	192 GB
Memory Bandwidth	900 GB/s	1,555 GB/s	2,039 GB/sec	3,352 GB/s	8,000 GB/sec
TDP	300 Watts	400 Watts	350 Watts	700 Watts	1,000 Watts
Transistors	21.1 B	54.2 B	80 B	80 B	208 B
GPU die size	815 mm^2	826 mm^2	814 mm^2	814 mm^2	–
Manufacturing	12 nm	7 nm	TSMC 4N	TSMC 4N	TSMC 4NP

various numerical formats, including FP16, TF32, FP32, FP64, among others [59]. The performance of 32-bit floating-point was much improved by the new TF32 tensor format, which at the time was just a tenth of the speed of the DL framework's V100 32-bit floating-point instruction. NVIDIA launched the H100 GPU in two variants in 2022: PCIe and SXM5. According to the specifications of Table 3, with tensor support, the H100 SXM5 outperforms the A100 by roughly 3.17 times in FP16 and TF32 performance, 3.4 times in FP64 performance, and roughly twice as fast in memory bandwidth.

In March 2024, NVIDIA went on to unveil its latest advancements in AI technology, including the Blackwell B200 GPU and the AI superchip, Blackwell GB200. Blackwell B200, featuring 208 billion transistors, is manufactured using a specially designed 4NP TSMC [63] process. This involves joining two GPU dies, limited by reticle, with a 10 TB/second chip-to-chip connection to form a single, unified GPU. This innovation has earned the chip the title of "World's Most Powerful Chip". This development is one of six groundbreaking technologies introduced with the Blackwell architecture, enabling real-time large language model (LLM) inference and AI training for models with up to 10 trillion parameters [64].

Through a 900 GB/s ultra-low-power NVLink chip-to-chip connection, the NVIDIA GB200 Grace Blackwell Superchip integrates two B200 Tensor Core GPUs with the Grace CPU [65]. Systems powered by the GB200 can connect to the newly introduced Quantum-X800 InfiniBand [66] and SpectrumTM-X800 [67] Ethernet platforms, designed to deliver optimal AI performance with advanced networking capabilities of up to 800 Gb/s [64].

The GB200 computing tray, based on the latest NVIDIA MGX architecture, incorporates four Blackwell GPUs and two Grace CPUs. It features NVLink connectors for the NVLink cable cartridge, PCIe Gen 6 compatibility for high-speed networking, and liquid cooling capabilities with cold plates and connections. Boasting 1.7 TB of fast memory and delivering 80 petaflops of AI performance, the GB200 compute tray represents a significant advancement in computing power.

NVIDIA plans to package the GB200 into a multi-chip ensemble called the GB200 NVL72 [68] superchip to meet the demanding needs of large enterprises and corporations. The GB200 NVL72, a crucial component of NVIDIA's advanced offerings, features the GB200-a multi-node, rack-scale, liquid-cooled system engineered for the most computationally demanding tasks. 36 Grace Blackwell Superchips, comprising 72 Blackwell GPUs and 36 Grace CPUs, are integrated into this system via fifth-generation NVLink.

Additionally, the GB200 NVL72 has NVIDIA BlueField-3 [69] data processing units, which allow the GPU to calculate cloud network acceleration, zero-trust security, composable storage, and elasticity in hyperscale AI clouds. The GB200 NVL72 achieves up to a 25x decrease in cost and energy consumption while providing up to a 30x performance boost for large language model (LLM) inference jobs when compared to an equivalent number of H100 Tensor Core GPUs [64].

2.4   Tensor Processing Unit

The Tensor Processing Unit (TPU) is an AI accelerator ASIC (application-specific integrated circuit) developed for neural network machine learning, utilizing Google's proprietary TensorFlow software. TPUs, which lack specialized hardware for texture mapping, are optimized to efficiently perform more input/output operations per joule and excel at managing large-scale computations using low-precision arithmetic, sometimes down to 8-bit, compared to GPUs [70].

The host processor utilizes CISC instructions to power the 8-bit matrix multiplication engine of the first-generation TPU, which is connected to a PCIe 3.0 bus. It is produced using a die size of $\leq 331\,mm^2$ with 28 nm technology. The TPU operates at a clock frequency of 700 MHz, with a thermal design power ranging from 28 to 40 W. It features 28 MiB of on-chip memory and 4 MiB of 32-bit accumulators, which process the output from a $256{\times}256$ systolic array composed of 8-bit multipliers [71]. Additionally, the TPU package includes eight gigabytes of dual-channel 2,133 MHz DDR3 SDRAM with a bandwidth of 34 GB/s. Instructions for the TPU encompass applying activation functions, executing matrix multiplications or convolutions, and managing data transfer to and from the host [71].

In May 2017, the second-generation TPU was introduced. The first-generation TPU design was limited by memory bandwidth, according to Google; however, the second-generation design achieved 45 teraFLOPS of performance and 600 GB/s of bandwidth by utilizing 16 GB of High Bandwidth Memory

[71]. After that, the TPUs were organized into four-chip modules, each delivering 180 teraFLOPS of performance [72]. Subsequently, 64 of these modules were combined to create 256-chip pods, achieving a performance of 11.5 petaFLOPS [72]. Notably, the second-generation TPUs introduced the ability to compute in floating point using the bfloat16 format developed by Google Brain, whereas the first-generation TPUs could only calculate in integers. This advancement enabled both training and inference of machine learning models on second-generation TPUs. According to Google, TensorFlow applications can utilize these second-generation TPUs on the Google Compute Engine [73].

On May 8, 2018, Google unveiled the third-generation TPU [74]. According to reports, these processors would be put in pods with four times as many chips as the previous generation and would be twice as powerful as the second-generation TPUs [75] [76]. This design allows each pod to support up to 1,024 chips, resulting in an eightfold performance improvement.

Tensor Processing Units (TPU v4) were unveiled by Google CEO Sundar Pichai in his keynote address at the virtual Google I/O conference on May 18, 2021. TPU v4 demonstrated performance improvements exceeding twice that of TPU v3. Pichai stated that "A single v4 pod comprises 4,096 v4 chips, each providing ten times greater interconnect bandwidth per chip at scale than any other existing networking technology" [77]. According to a Google document from April 2023, TPU v4 outperforms A100 by 5–87% on machine learning benchmarks [78].

Google said in 2021 that a fresh deep reinforcement learning application is powering the TPU v5's physical architecture [79]. Considering Google's statement that TPU v5 nearly doubles the performance of TPU v4, combined with TPU v4's previously reported advantages over Nvidia's A100, some speculate that TPU v5 could match or even surpass the capabilities of Nvidia's H100 [80]. The budget-friendly TPU v5e, analogous to the lighter v4i version of the previous generation, is included in Google's fifth-generation TPUs. Additionally, Google introduced the TPU v5p in December 2023, claiming performance competitive with NVIDIA's H100 GPU [81].

Google unveiled the TPU v6 during the Google I/O conference in May 2024; it will be available later that year. Google reported a 4.7-fold increase in performance over the TPU v5e [82], attributed to higher clock speeds and larger matrix multiply units. Additionally, both the capacity and bandwidth of High Bandwidth Memory (HBM) have been doubled. Up to 256 Trillium units can be housed within a pod [83].

2.5   Cloud-Based AI

Currently, AI servers are becoming an essential part of the IT infrastructure for many organizations and businesses worldwide. With the explosion of artificial intelligence applications, the demand for powerful and efficient servers to process complex AI models is increasing significantly. Major technology companies such as Google, Amazon, Microsoft, and NVIDIA are heavily investing in the development and deployment of advanced AI servers.

AI experienced a significant breakthrough with the official release of Chat-GPT on November 30, 2022. Powered by large language models (LLMs), Chat-GPT allows users to refine and steer conversations to achieve a desired length, format, style, level of detail, and language. This iterative approach, where successive prompts and replies build on the previous context, has revolutionized human-AI interaction. As a result, ChatGPT has become a versatile and responsive tool, capable of catering to a wide range of user needs and applications. This paper addresses the question: What hardware supports the operation of Chat-GPT? Specifically, what hardware was employed for its training and inference?

In May 2020, Microsoft unveiled a new supercomputer designed exclusively for OpenAI to train GPT-3, the predecessor to the machine learning model utilized for ChatGPT (GPT-3.5). This supercomputer was equipped with over 285,000 CPU cores, 10,000 NVIDIA V100 GPUs, and provided 400 Gb/s of network connectivity for each GPU server, supporting a model with 175 billion parameters [84]. ChatGPT was fine-tuned from the GPT-3.5 series, which completed its training in early 2022. Both ChatGPT and GPT-3.5 were trained on Azure AI's supercomputer infrastructure, leveraging a substantial number of CPU cores and NVIDIA A100 GPUs [85].

Moreover, ChatGPT-4, Google Gemini, and many sophisticated AI servers already operate globally, providing customers with reliable AI services through a combination of state-of-the-art technological solutions and high-performance hardware. These servers offer stringent security protections, flexible scalability, and the capacity to handle and analyze massive volumes of data. As hardware technology continues to advance, AI servers are becoming indispensable platforms for implementing complex artificial intelligence applications and quickly meeting the growing demands of the market.

3 Edge AI

3.1 Origins and Development of Edge AI

Today, leading technology companies are racing to develop hardware to meet the growing demands of AI. As algorithms become more complex and the volume of data increases, the development of cloud-based AI is becoming increasingly crucial. However, cloud-based AI still faces many limitations in real-time applications. While super-computing devices are equipped with immense AI processing capabilities, the transmission (including issues related to bandwidth, latency, etc.) from the cloud infrastructure to user devices remains a significant challenge. Because AI models are stored and processed on cloud-based AI services, their use entails a dependency on third parties. This dependency poses potential risks, including data loss and privacy breaches. Edge computing has emerged to address the limitations of cloud computing, particularly regarding latency, bandwidth, and privacy. Notable applications of edge computing that garner significant interest and research include security, mobile networks, healthcare, imaging, voice, and military applications. Since 2018, extensive research and development have focused on edge AI. Edge AI is used to predict diseases such as respiratory diseases [86] and chronic obstructive pulmonary disease (COPD) [87]. In the study by HH Kumar et al., the classical K-Means algorithm was proposed to enhance privacy in edge AI [88]. Additionally, Jiwei Zhang et al. developed Anticoncealer, a system capable of detecting malware and dangerous behaviors within IoT networks. This system utilizes edge AI to address the inefficiencies of IoT security protocols [89]. Author Xiaofei Wang et al. in [90] have surveyed and compared the use of deep learning in edge AI and cloud-based AI. Three widely discussed and popular terms in edge AI are neural network model optimization, AI on mobile devices, and AI on embedded devices. In the context of neural network model optimization, the two most popular frameworks, TensorFlow and PyTorch, will be examined. For AI on mobile devices, the focus will be on the two major operating systems: IOS and Android. Regarding AI on embedded devices, a deeper exploration of TinyML [91] will be conducted. Additionally, there are studies that have applied edge AI for continuous monitoring of vital signs such as body temperature, heart rate, and respiratory rate for early disease prediction and diagnosis. The authors employed optical methods, including RGB cameras and thermal imaging systems, which are non-contact and highly valuable in pandemic scenarios such as COVID-19. Supervised learning models operating on edge devices like the Raspberry PI 4 achieved an accuracy of 98.97% and reliability of 98.4% for the RGB model, and an accuracy of 96.7% and reliability of 95.18% for the thermal model [92].

3.2 Edge AI Hardware

Edge AI hardware refers to hardware devices designed to deploy and run artificial intelligence (AI) applications directly at the network "edge" meaning closer to the data source or end-users rather than processing data in a centralized data center or cloud. Edge AI hardware aims to improve security, lower bandwidth consumption, deliver AI services with minimal latency, and allow for continued operation even when there isn't a reliable network connection (Table 4).

Mobile AI Chips are processors integrated into mobile devices such as smartphones, tablets, and wearables, optimized to handle tasks related to artificial intelligence (AI) and machine learning. These chips provide the necessary computational power to perform complex AI tasks directly on the device, enhancing performance, reducing latency, and saving bandwidth compared to cloud-based data processing. Authors Tianxiang Tan and Guohong Cao have discussed and improved the performance of deep neural networks (DNNs) on mobile devices using neural processing units (NPUs). Specifically, the authors used the Huawei Mate 10 Pro, an Android operating system smartphone, to conduct their experiments. Tools such as the Caffe deep learning framework and Huawei DDK were employed to optimize DNN models partitioned across the CPU and NPU. The results indicated that while NPUs can reduce processing time compared to CPUs, accuracy issues arise due to the support for only low-precision computations (FP16 or FP8 instead of FP32). The paper proposes model partitioning techniques to maximize NPU performance without significantly compromising

Table 4. Characteristics of AI chips utilized in edge computing

Name	Category	Performance	Model	Application
Coral Dev Board TPU [93]	ASIC	4 TOPS	EfficientNet-EdgeTPU, MobileNet v2, etc.	Human Emotion Analysis on Mobile Devices, Cancer Diagnosis, Plant Leaf Disease Classification, etc.
NVIDIA Jetson Nano [94]	GPU	472 GFLOPS	TensorFlow, PyTorch, Caffe/Caffe2, Keras, MXNet, etc.	Edge Camera, Warehouse Robot, Digital Signage, etc.
Qualcomm Hexagon NPU (Snapdragon X Plus) [95]	NPU	45 TOPS	HRNet, MobileNetV2, BERT(GLUE), etc.	Mobile Device Task.
BeagleBone AI [96]	SBC	-	TIDL API, OpenCL Runtime	Industrial Automation, Smart Cameras, Robotics, etc.
Intel Core Ultra 9 Processor 288V [97]	CPU	120 TOPS	OpenVINO^{TM}, WindowsML, DirectML, ONNX RT, WebNN, etc.	General Task.
Halio 8 [98]	ASIC	26 TOPS	MobileNet v1 SSD, Mono Depth 2, etc.	Automotive, Advanced Driver Assistance Systems, Smart Surveillance Cameras, Patient Monitoring Systems, etc.

accuracy [99]. In addition, major companies like Apple have introduced their A-series chips with Neural Engine since 2017 for iPhone X and later models. The Neural Engine plays a crucial role in driving AI features on Apple devices, from enhancing photo quality to improving security features and overall performance. It is an integral part of Apple's strategy to enhance user experience through AI technology [100]. Leading smartphone chip manufacturers like Qualcomm have also made significant advancements with their Snapdragon chips featuring the Hexagon DSP. The Hexagon DSP is a specialized processor designed to handle complex operations, particularly effective for tasks such as audio and video decoding. It includes features like Hexagon Vector Extensions (HVX), which enable efficient vector operations crucial for handling large data sets and parallel processing [101]. The Hexagon DSP has evolved through many generations, with the latest versions developed by Qualcomm, recently announced in the Snapdragon X lite and Snapdragon X plus chip series. According to the developer, these chips integrate the Qualcomm Hexagon NPU with performance reaching up to 45 TOPS, delivering significant improvements in processing power and energy efficiency. These DSPs support a wide range of applications, from basic audio decoding to complex machine learning inference, integrating functions between the CPU and GPU by combining vector and tensor units with capabilities like branch prediction and multithreading [102].

AI acceleration devices are specialized hardware designed to speed up artificial intelligence tasks, particularly machine learning and deep learning models. These units are optimized to handle AI workloads more efficiently than traditional processors like CPUs (Central Processing Units). Some common specialized AI acceleration devices include:

Application-Specific Integrated Circuit (ASIC): ASICs are chips designed specifically for a particular application. In the realm of edge AI, ASICs play a crucial role due to their ability to deliver high performance with low power consumption, which is essential for AI applications on edge devices such as IoT devices, smart cameras, and mobile devices. A prime example of an ASIC designed for edge applications is Google's Edge TPU. Google has developed compact, energy-efficient TPU versions for edge applications. The Edge TPU can be utilized in devices like Google Coral to run deep learning models directly on the device (Fig. 5).

Fig. 5. Coral Dev Board TPU [103]

The Coral Dev Board TPU, developed by Google, is built on the NXP i.MX 8M SoC configuration. It features a quad-core Cortex-A53 CPU with a maximum clock speed of 1.5 GHz, a GC 7000 Lite GPU supporting OpenGL, OpenCL, and Vulkan, and a Video Processing Unit for video encoding and decoding. Alongside its AI accelerator delivering 4 TOPS of performance, and direct support for TensorFlow Lite, the Coral Dev Board TPU is a powerful tool for AI developers.

It provides high-performance deep learning capabilities directly on edge devices while maintaining low power consumption [93]. With such a robust configuration, there have been significant experiments aimed at embedding powerful AI models into edge devices. Author Brendan Reidy and his colleagues experimented with the Coral Dev Board TPU to research and deploy Transformer models, which are highly powerful but resource-intensive, to be effectively deployed on edge TPUs designed to optimize performance and energy for AI applications at the edge. The paper conducted practical experiments to measure the performance of Transformer models when running on edge TPU. These models include variants like BERT and GPT, customized to fit the resource constraints of the edge TPU. The experiments involved comparing processing speed, energy consumption, and accuracy of the models running on edge TPU versus other platforms such as CPUs and GPUs. The results from these experiments provide insights into the practical capabilities of edge TPU in running Transformer models, including factors like inference time, energy consumption, and accuracy retention when transitioning from more powerful platforms to edge TPU [104].

FPGA (Field-Programmable Gate Array): is a type of flexible integrated circuit that can be reprogrammed after manufacturing to perform various tasks. In the context of edge AI, FPGAs play a crucial role due to their configurable nature and high performance, particularly in applications requiring real-time processing and specialized customization. Research on FPGAs focuses on leveraging their flexibility and performance to address challenges in deploying AI models on edge devices. Author Hadnagy and colleagues have explored effective methods for implementing Convolutional Neural Networks (CNNs) on FPGAs, using optimization techniques to enhance performance and reduce energy consumption [105].

SoC (System on a Chip): is an integrated circuit that combines all the necessary components of a computer system or other electronic systems onto a single chip. SoCs typically integrate multiple components, including CPU, GPU, NPU (Neural Processing Unit), and other supporting hardware. This allows for performing various AI functions and data processing directly at the edge device without relying on cloud services. SoCs are designed to optimize performance and energy efficiency. The components within an SoC can operate synchronously to enhance processing efficiency and reduce power consumption, which is crucial for edge devices. SoCs help reduce the size and cost of devices by consolidating multiple functions onto a single chip, as opposed to using multiple separate integrated circuits. Current research on SoCs focuses on integrating hardware components such as CPU, GPU, NPU, and other support circuits into a single SoC to optimize performance for AI applications.

Embedded AI Systems: These systems comprise embedded devices designed to perform specific AI tasks. Within the context of edge AI, these embedded systems play a crucial role in providing real-time analytics and information processing capabilities for various applications. The ability of Embedded AI Systems to process data at the network edge reduces latency and increases response speed, which is vital for real-time applications such as autonomous vehicles and industrial robots. By processing data on the device, embedded AI systems minimize security and privacy risks as data does not need to be transmitted to the cloud.

Single-Board Computers (SBCs): play a significant role in edge AI applications due to their ability to provide a flexible, cost-effective, and easily deployable platform. One of the most popular SBCs in AI projects is the Raspberry Pi. The advantages offered by the Raspberry PI include robust performance, a large developer community, and extensive software support. Combined with the ability to connect multiple sensors and AI modules, it has facilitated substantial research in the development of edge AI devices. Author Mahmut Taha Yazici used Raspberry Pi devices to deploy and execute machine learning models directly on edge devices. The results indicated that applying AI algorithms at the edge significantly reduces network latency and bandwidth [106].

BeagleBone AI: is a new, sophisticated, and advanced single-board computer (SBC). It is designed to bridge the gap between conventional SBCs and powerful industrial computers. Targeted at machine learning and computer vision applications, the BeagleBone AI uses the Texas Instruments AM5729 SoC and is highly versatile with its processors and accelerators, supporting a wide range of applications. It is compatible with all major machine learning frameworks such as TensorFlow, Caffe, and MXNET. The BeagleBone AI can efficiently run machine learning algorithms by leveraging its IPU and GPU accelerators. It also features a specialized EVE system for running computer vision applications. BeagleBone recommends using the Cloud9 online IDE for programming.

Besides SBC embedded devices, the NVIDIA Jetson Nano is a popular edge AI platform, providing a powerful and compact solution for running artificial intelligence (AI) applications at the edge. Jetson Nano is equipped with a Maxwell GPU with 128 cores, delivering significant processing power for AI tasks. Additionally, Jetson Nano operates on NVIDIA's JetPack SDK, which includes AI development tools and libraries such as TensorRT, CUDA, and cuDNN. These tools enhance Jetson Nano's capability to deploy edge AI devices effectively [107].

3.3   Edge AI Algorithm

Smartphones, Internet of Things (IoT) devices, and other edge devices are often constrained by limited processing power and battery life. Therefore, Algorithms that facilitate on-device real-time processing and decision-making are essential, as this eliminates the need to transmit data to centralized servers or the cloud for analysis. Techniques such as model compression, quantization, and efficient neural network architectures are frequently studied to optimize resources. Notable edge AI algorithms that have attracted significant interest include:

The primary goal of TinyML [108] is to implement machine learning models on small embedded devices with constrained memory and energy sources. By enabling the execution of sophisticated algorithms without requiring powerful computing systems, TinyML optimizes performance on small devices and conserves energy. The components of TinyML include libraries, software, hardware, and optimization methods in Fig. 6. Regarding hardware, Tiny devices are geared towards compact devices capable of connecting with various surrounding devices, including IoT-supporting devices such as Embedded AI systems, Mobile AI Chips, ASICs, and FPGAs.

These devices need to be designed with minimal energy consumption while still possessing sufficient computational capacity to perform machine learning tasks. TinyML is integrated with software tools and support libraries such as TensorFlow Lite, uTensor, and CMSIS-NN, which are deployed in stable environments like Embedded systems to optimize machine learning models for embedded devices. TinyML employs approaches including model compression, quantization, and constrained neural architecture search to make machine learning models suitable for resource-limited devices. TinyML is gaining significant interest and is being extensively applied across various domains, such as voice-activated devices that can recognize and respond to user commands, object recognition and classification in images and videos, and translation of sign language into audio or text.

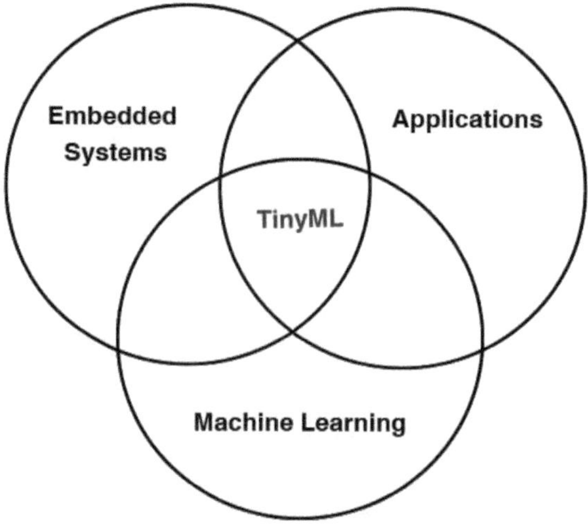

Fig. 6. Composition of TinyML

A set of deep neural network architectures called MobileNet was created specifically to execute machine learning applications on mobile and edge devices with constrained memory and processing capacity. MobileNet has grown in popularity as an edge AI Algorithm option due to its ability to offer a good balance between accuracy and resource efficiency. MobileNet uses a method called depthwise separable convolutions to minimize the number of parameters and operations needed compared to conventional neural network models. This method divides a convolutional layer into two sections: pointwise and depthwise convolution, as shown in Fig. 7. This considerably lowers computing costs without significantly sacrificing model accuracy. The width multiplier, a parameter in MobileNet, allows for adjusting the model size by reducing the number of channels in the network, thereby lowering resource requirements. MobileNet reduces not only the size of the model but also the computational costs by allowing modifications to the input resolution. This is achieved through the resolution multiplier parameter.

In the realm of computer vision and artificial intelligence, YOLO (You Only Look Once) is a well-known object identification technique created for real-time object recognition. It is a popular choice in edge AI systems where quick processing times and efficient resource use are essential. Unlike traditional methods that analyze distinct areas of an image independently, YOLO completes object detection in a single neural network run. This allows YOLO to operate at high speeds, making it ideal for real-time detection applications such as security cameras, self-driving cars, and Internet of Things (IoT) devices.

The YOLO architecture partitions the input image into a grid. For each resulting grid cell, the model predicts bounding boxes and associated class probabilities. This strategy facilitates the simultaneous determination of both an object's classification and its spatial location within the image. Designed to minimize computational requirements, YOLO can operate on embedded systems and smartphones with limited computing power, making it an optimal choice for edge AI Algorithms. Due to its efficiency, speed, and ability to provide rapid and accurate object recognition, YOLO is one of the top choices for object detection and tracking tasks in edge AI applications.

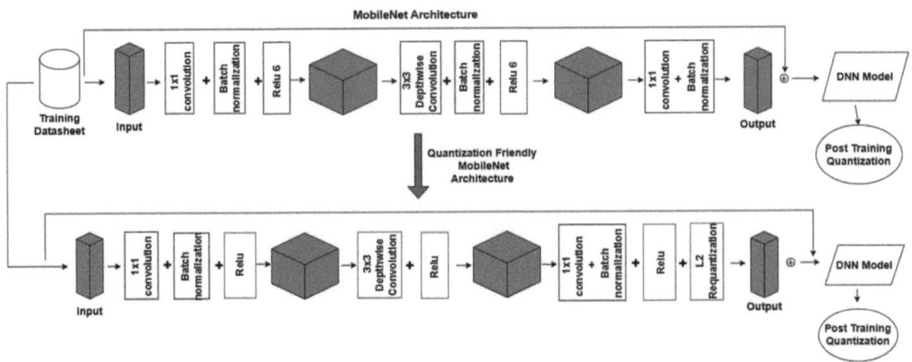

Fig. 7. MobileNet Architecture

4   Discussion

4.1   Edge AI and Cloud-Based AI

Edge and cloud-based AI represent two distinct approaches to artificial intelligence processing. Edge AI performs computations directly on the device, offering minimal latency and enhanced data privacy, making it ideal for time-sensitive applications. Edge AI is expected to expand at a compound annual growth rate (CAGR) of 21% between 2023 and 2030 [109]. In contrast, cloud-based AI leverages the vast computational.

Power of centralized data centers, enabling complex and scalable data analysis. With a compound annual growth rate (CAGR) of 30.9% during the same time frame, the cloud-based AI market is anticipated to rise even more quickly [110]. North America currently leads the cloud-based AI market. Each technology has its strengths: edge AI excels in providing real-time, on-device intelligence, while cloud-based AI offers powerful remote computing capabilities. Table 5 below provides a comparative overview of their key characteristics.

Latency and Bandwidth: for cloud-based AI, applications that are sensitive to latency and bandwidth, such as real-time applications. This is a big problem with cloud-based systems, especially in situations where milliseconds can matter. Instead of sending data to the cloud and waiting for a response, edge computing with AI integration processes data locally on the device or on a nearby server, reducing latency and bandwidth. Stores that use AI loss prevention systems, for instance, are unable to wait for a response from the system for more than a few seconds. To recognize and handle anomalous conduct right away, they want instant alerts. Similarly, disaggregate object counting automation in industrial manufacturing with edge AI is shown in Fig. 8. It has the ability to quickly process images, recognize objects, and count in real time without needing to send data to a remote server, which helps save time.

Data Security and Privacy: This remains a major concern in cloud-based AI. While cloud providers implement robust security measures, including machine learning-based threat detection, the transfer of sensitive data to and from the cloud introduces inherent risks. Data breaches, leaks, or attacks can compromise user privacy. Edge AI addresses these concerns by keeping data local, either on the device or a private network [112]. This significantly reduces exposure to external threats, making it well-suited for handling sensitive information like healthcare records or financial data.

Reliability and Internet Connectivity: Cloud-based AI and data processing heavily rely on a stable internet connection. Poor network connectivity can severely impact real-time applications and machine learning processes, particularly in remote or in areas with unreliable internet. A significant portion of the global population still lacks internet access [113], highlighting this challenge. In contrast, edge AI operates independently of network connectivity, offering greater resilience and reliability. Even during network outages, edge AI systems can continue to function, ensuring minimal disruption to operations.

Table 5. Comparison between Edge AI and Cloud-based AI

Criterion	Edge AI	Cloud-based AI
Latency and bandwidth	Low latency	High latency
Data security and privacy	Enhanced	Low
Reliability and connectivity	At the edge of the local network	Within the internet
Scalability and accessibility	Limited local hardware	Highly scalable
Cost	Higher initialization, lower operation	Lower initialization, high service fees
Data consistency and standardization	Different manufacturers, conflict standards	Homogeneous and standardized
Deployment	Individual devices	Centralized deployment
Use cases	Real-time processing	Complex data analysis

Fig. 8. Object counting in manufacturing processes with edge AI development [111]

Scalability and Accessibility: Edge AI devices generally possess constrained computational capabilities [114]. The pruning algorithm is important to optimize computation because hardware scalability is constantly constrained on edge AI devices. Over-pruning might lead to the performance decrease [115]. In contrast to the constraints of edge AI, linked to scalability and flexibility, cloud-based AI displays effectiveness in satisfying this demand. Cloud-based AI can readily scale to accommodate growing data requirements and increasingly complicated applications without the need to replace hardware. For example, a huge retail company expects a significant rise in online purchases during the holiday season. By implementing cloud computing combined with artificial intelligence (AI), enterprises may instantly assign additional computing resources to accommodate surges in demand, assuring uninterrupted service.

Cost: Cloud-based AI often boasts lower initial costs but ongoing expenses for data transfer, storage, and processing can escalate over time. Cloud infrastructure is a rapidly growing expense for many businesses [116]. Edge AI, while requiring higher upfront investment in devices, can reduce long-term costs by minimizing data transfer, especially in scenarios where connectivity is expensive.

Data Consistency and Standardization: AI Edge devices come from many different manufacturers. Making sure they all "speak the same language" is important but can be a concern [117]. While this decentralized technique provides benefits, such as lower latency, it also poses obstacles. As shown in Fig. 9

the edge AI projects and ideas that have been discussed in scholarly endeavors and standardization efforts have been underway since 2009 [118]. Cloud-based AI, with its centralized data management, offers inherent advantages in maintaining data consistency and standardization.

Deployment: Edge AI deploys directly on local devices or nearby servers, enabling rapid processing and low latency. Cloud-based AI, on the other hand, is deployed in centralized data centers and excels at handling computationally intensive tasks like deep learning and large-scale data analytics. The optimal deployment strategy depends on the specific application requirements.

Use Cases: Cloud-based AI is often preferred for complex data analysis and model training, while edge AI shines in real-time applications where immediate responses are critical. Edge AI's ability to function even with intermittent or no internet connectivity makes it suitable for various scenarios.

This section has offered an outline of the pros and cons of edge AI and cloud-based AI. From there, it helps customers choose the right infrastructure, which might include edge AI, cloud-based AI, or a combination of both to construct their apps.

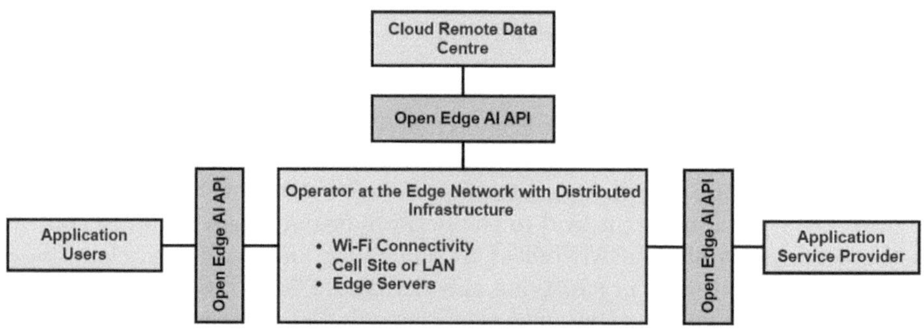

Fig. 9. Standardization initiatives for edge AI development [119]

5 Conclusion

The rapid evolution of AI hardware has fueled significant advancements in cloud-based and edge AI, each with its unique strengths and applications. Cloud AI, powered by high-performance chips, excels at handling massive datasets and complex models, making it ideal for large-scale data analytics and scientific research tasks. Edge AI, on the other hand, leverages specialized hardware to enable real-time decision-making at the source of data, critical for applications like autonomous vehicles and smart city infrastructure.

A balanced approach, recognizing the complementary strengths of both cloud and edge AI, is essential for the future of AI. By understanding the specific requirements of different applications and leveraging the right combination of cloud and edge capabilities, we can create more intelligent, responsive, and scalable solutions. As AI hardware continues to evolve, the role of edge AI will only grow, driving innovation and progress across various industries. The integration of cloud and edge AI technologies plays a pivotal role in fully harnessing the power of artificial intelligence to address real-world problems and contribute to building a more sustainable and intelligent future.

Acknowledgment. The authors would like to thank the VNU University of Engineering and Technology for supporting this work.

References

1. MarketsandMarkets: artificial intelligence (AI) market by offering (discriminative AI, generative AI, hardware, services), technology (ml, nlp, context-aware ai, computer vision), business function (marketing & sales, HR), vertical and region - global forecast to 2030. https://www.marketsandmarkets.com/Market-Reports/artificial-intelligence-market-74851580.html, Accessed 27 July 2024
2. Research, G.V.: Artificial intelligence market to reach \$1,811.75 billion by 2030. https://www.grandviewresearch.com/press-release/global-artificial-intelligence-ai-market, Accessed 28 July 2024
3. Xiong, Z., Zhang, Y., Niyato, D., Wang, P., Han, Z.: When mobile blockchain meets edge computing. IEEE Commun. Magazine **56**(8), 33–39 (2018)
4. Howarth, J.: 57 new artificial intelligence statistics (2024). https://explodingtopics.com/blog/ai-statistics, Accessed 28 July 2024
5. Bekker, A.: Big data in manufacturing: use cases + guide on how to start. https://www.scnsoft.com/data/big-data-in-manufacturing-use-cases, Accessed 28 July 2024
6. Toolify: Unleashing AI in unilever: a HR case study. https://www.toolify.ai/ai-news/unleashing-ai-in-unilever-a-hr-case-study-450814, Accessed 28 July 2024
7. Moon, M.: Chatgpt reportedly reached 100 million users in January. https://www.engadget.com/chatgpt-100-million-users-january-130619073.html, Accessed 28 July 2024
8. Meer, D.V.: Number of ChatGPT users and key stats (2024). https://www.namepepper.com/chatgpt-users, Accessed 25 July 2024
9. Duarte, F.: Number of ChatGPT users (2024). https://explodingtopics.com/blog/chatgpt-users, Accessed 25 July 2024
10. Dean, J., et al.: Large scale distributed deep networks. In: NIPS (2012)
11. Armbrust, M., et al.: A view of cloud computing. CACM, 50–58 (2010)
12. Shi, W., Cao, J., Zhang, Q., Li, Y., Xu, L.: Edge computing: vision and challenges. IEEE Internet Things J. **3**(5), 637–646 (2016)
13. Zhang, Q., Yang, L.T., Chen, Z., Li, P.: A survey on deep learning for big data. Inf. Fusion **42**, 146–157 (2018). https://www.sciencedirect.com/science/article/pii/S1566253517305328
14. Satyanarayanan, M.: The emergence of edge computing. Computer **50**(1), 30–39 (2017)
15. Turing, A.M.: Computing machinery and intelligence. Springer (2009)

16. McCarthy, J., Minsky, M.L., Rochester, N., Shannon, C.E.: A proposal for the dartmouth summer research project on artificial intelligence, august 31, 1955. AI Mag. **27**(4), 12 (2006)
17. Newell, A., Shaw, J.C., Simon, H.A.: Report on a general problem solving program. In: IFIP Congress. vol. 256, p. 64. Pittsburgh (1959)
18. Weizenbaum, J.: Eliza—a computer program for the study of natural language communication between man and machine. Commun. ACM **9**(1), 36–45 (1966)
19. Rosenblatt, F.: The perceptron: a probabilistic model for information storage and organization in the brain. Psychol. Rev. **65**(6), 386 (1958)
20. Giacaglia, G.: Making things think: how AI and deep learning power the products we use. Holloway, Incorporated (2021)
21. Nikolopoulos, C.: Expert systems: introduction to first and second generation and hybrid knowledge based systems. CRC Press (1997)
22. Crevier, D.: AI: the tumultuous search for artificial intelligence 1993 new york (1993)
23. McKinzie, W.R.: The fifth generation. Proc. IEEE **73**(3), 493–494 (1985)
24. Rumelhart, D.E., Hinton, G.E., Williams, R.J.: Learning representations by back-propagating errors. Nature **323**(6088), 533–536 (1986)
25. IBM: Ibm's computer checkmated a human chess champion in a computing tour de force. https://www.ibm.com/history/deep-blue
26. Hinton, G., et al.: Deep neural networks for acoustic modeling in speech recognition: the shared views of four research groups. IEEE Sign. Process. Magazine **29**(6), 82–97 (2012)
27. LeCun, Y., Bengio, Y., Hinton, G.: Deep learning. Nature **521**, 436–44 (05 2015)
28. Vaswani, A., et al.: Attention is all you need. In: Guyon, I., et al., (eds.) Advances in Neural Information Processing Systems. vol. 30. Curran Associates, Inc. (2017)
29. Corbett, M., Sajal, S.: AI in cybersecurity. In: 2023 Intermountain Engineering, Technology and Computing (IETC), pp. 334–338 (2023)
30. Ma, Y., Wang, Z., Yang, H., Yang, L.: Artificial intelligence applications in the development of autonomous vehicles: a survey. IEEE/CAA J. Automatica Sinica **7**(2), 315–329 (2020)
31. DeepMind: Imagen 3. https://deepmind.google/technologies/imagen-3/
32. DeepMind: Alphafold. https://deepmind.google/technologies/alphafold/
33. Aspray, W.: The intel 4004 microprocessor: what constituted invention? IEEE Ann. History Comput. **19**(3), 4–15 (1997)
34. Morse, R.: Mazor, pohiman: intel microprocessors— 8008 to 8086. Computer **13**(10), 42–60 (1980)
35. The IBM pc. https://www.ibm.com/history/personal-computer
36. Patterson, D.A., Ditzel, D.R.: The case for the reduced instruction set computer. ACM SIGARCH Comput. Archit. News **8**(6), 25–33 (1980)
37. Patterson, D.A., Hennessy, J.L.: Computer organization and design ARM edition: the hardware software interface. Morgan Kaufmann (2016)
38. Hennessy, J., et al.: Mips: a microprocessor architecture. ACM SIGMICRO Newsletter **13**(4), 17–22 (1982)
39. Gelsinger, P.: Microprocessors for the new millennium: challenges, opportunities, and new frontiers. In: 2001 IEEE International Solid-State Circuits Conference. Digest of Technical Papers. ISSCC (Cat. No.01CH37177), pp. 22–25 (2001)
40. Gepner, P., Kowalik, M.: Multi-core processors: new way to achieve high system performance. In: International Symposium on Parallel Computing in Electrical Engineering (PARELEC'06), pp. 9–13 (2006)

41. Intel: Intel core 2 quad q6600. https://www.intel.com/content/www/us/en/products/sku/29765/intel-core2-quad-processor-q6600-8m-cache-2-40-ghz-1066-mhz-fsb/specifications.html?wapkw=intel%20core%202%20quad%20q6600

42. Intel: Intel core i7-7700. https://www.intel.com/content/www/us/en/products/sku/97128/intel-core-i77700-processor-8m-cache-up-to-4-20-ghz/specifications.html

43. AMD: Amd ryzen 7 7600x. https://www.amd.com/en/products/processors/desktops/ryzen/7000-series/amd-ryzen-5-7600x.html

44. AMD: Amd ryzen 9 9950x. https://www.amd.com/en/products/processors/desktops/ryzen/9000-series/amd-ryzen-9-9950x.html#

45. Intel: Intel core i9 14900ks. https://www.intel.com/content/www/us/en/products/sku/237504/intel-core-i9-processor-14900ks-36m-cache-up-to-6-20-ghz/specifications.html

46. Moore, G.: Moore's law. Electron. Magazine **38**(8), 114 (1965)

47. Khan, F.H., Pasha, M.A., Masud, S.: Advancements in microprocessor architecture for ubiquitous AI– an overview on history, evolution, and upcoming challenges in AI implementation. Micromachines **12**(6) (2021)

48. Luebke, D., et al.: GPGPU: general purpose computation on graphics hardware (2004)

49. Wang, Y., et al.: Benchmarking the performance and energy efficiency of AI accelerators for AI training. In: 2020 20th IEEE/ACM International Symposium on Cluster, Cloud and Internet Computing (CCGRID), pp. 744–751 (2020)

50. Dally, W.J., Keckler, S.W., Kirk, D.B.: Evolution of the graphics processing unit (GPU). IEEE Micro **41**(6), 42–51 (2021)

51. Baji, T.: GPU: the biggest key processor for AI and parallel processing. In: Takehisa, K. (ed.) Photomask Japan 2017: XXIV Symposium on Photomask and Next-Generation Lithography Mask Technology. vol. 10454, p. 1045406. International Society for Optics and Photonics, SPIE (2017)

52. Ghorpade, J., Parande, J., Kulkarni, M., Bawaskar, A.: GPGPU processing in CUDA architecture. arXiv preprint arXiv:1202.4347 (2012)

53. Chetlur, S., et al.: CUDNN: efficient primitives for deep learning (2014)

54. Paszke, A., et al.: Automatic differentiation in pytorch (2017)

55. Vedaldi, A., et al.: Convolutional architecture for fast feature embedding. Cornell University (2014)

56. Abadi, M., et al.: Tensorflow: large-scale machine learning on heterogeneous systems (2015)

57. NVIDIA: Nvidia cuda-x libraries. https://developer.nvidia.com/gpu-accelerated-libraries, Accessed 19 July 2024

58. NVIDIA: Nvidia tesla v100 GPU architecture. https://images.nvidia.com/content/volta-architecture/pdf/volta-architecture-whitepaper.pdf, Accessed 19 July 2024

59. NVIDIA: Nvidia a100 tensor core GPU architecture. https://images.nvidia.com/aem-dam/en-zz/Solutions/data-center/nvidia-ampere-architecture-whitepaper.pdf, Accessed 19 July 2024

60. NVIDIA: Nvidia h100 tensor core GPU architecture. https://resources.nvidia.com/en-us-tensor-core?ncid=no-ncid, Accessed 19 July 2024

61. Smith, R.: Nvidia blackwell architecture and b200/b100 accelerators announced: going bigger with smaller data. https://www.anandtech.com/show/21310/nvidia-blackwell-architecture-and-b200b100-accelerators-announced-going-bigger-with-smaller-data, Accessed 28 July 2024

62. Svedin, M., Chien, S., Chikafa, G., Jansson, N., Podobas, A.: Benchmarking the NVIDIA GPU lineage. CoRR (2021)
63. AMAX: Tsmc's 4nm process powers nvidia's blackwell architecture GPU. https://www.trendforce.com/news/2024/03/19/news-tsmcs-4nm-process-powers-nvidias-blackwell-architecture-gpu-ai-performance-surpasses-previous-generations-by-multiples/, Accessed 28 July 2024
64. NVIDIA: Nvidia blackwell platform arrives to power a new era of computing. https://nvidianews.nvidia.com/news/nvidia-blackwell-platform-arrives-to-power-a-new-era-of-computing, Accessed 19 July 2024
65. NVIDIA: Nvidia grace CPU superchip. https://resources.nvidia.com/en-us-grace-cpu/data-center-datasheet?ncid=no-ncid, Accessed 28 July 2024
66. NVIDIA: Nvidia quantum-x800 infiniband platform. https://www.nvidia.com/en-us/networking/products/infiniband/quantum-x800/, Accessed 28 July 2024
67. AMAX: Nvidia spectrum-x networking platform. https://www.amax.com/product/nvidia-spectrum-x800-ethernet/, Accessed 28 July 2024
68. NVIDIA: Nvidia gb200 nvl72. https://www.nvidia.com/en-us/data-center/gb200-nvl72/, Accessed 28 July 2024
69. NVIDIA: Nvidia bluefield networking platform. https://www.nvidia.com/en-us/networking/products/data-processing-unit/, Accessed 28 July 2024
70. GOOGLE: Google supercharges machine learning tasks with tpu custom chip. https://cloud.google.com/blog/products/ai-machine-learning/google-supercharges-machine-learning-tasks-with-custom-chip, Accessed 05 Aug 2024
71. Jouppi, N.P., et al.: In-datacenter performance analysis of a tensor processing unit. In: Proceedings of the 44th Annual International Symposium on Computer Architecture, pp. 1–12 (2017)
72. Staff, A.: Google brings 45 teraflops tensor flow processors to its compute cloud. https://arstechnica.com/information-technology/2017/05/google-brings-45-teraflops-tensor-flow-processors-to-its-compute-cloud/, Accessed 04 Aug 2024
73. Kennedy, P.: Google cloud TPU details revealed. https://www.servethehome.com/google-cloud-tpu-details-revealed/ Accessed 04 Aug 2024
74. Frumusanu, A.: Google i/o 2018 opening keynote live-blog (10am pt). https://www.anandtech.com/show/12726/google-io-keynote-liveblog-10am-pt, Accessed 04 Aug 2024
75. Feldman, M.: Google offers glimpse of third-generation TPU processor. https://www.top500.org/news/google-offers-glimpse-of-third-generation-tpu-processor/ Accessed 04 Aug 2024
76. Teich, P.: Tearing apart google's TPU 3.0 AI coprocessor. https://www.nextplatform.com/2018/05/10/tearing-apart-googles-tpu-3-0-ai-coprocessor/, Accessed 04 Aug 2024
77. Weiss, T.R.: Google launches TPU V4 AI chips. https://www.hpcwire.com/2021/05/20/google-launches-tpu-v4-ai-chips/ Accessed 04 Aug 2024
78. Jouppi, N., et al.: TPU v4: an optically reconfigurable supercomputer for machine learning with hardware support for embeddings. In: Proceedings of the 50th Annual International Symposium on Computer Architecture, pp. 1–14 (2023)
79. Mirhoseini, A., et al.: A graph placement methodology for fast chip design. Nature **594**(7862), 207–212 (2021)
80. Afifi-Sabet, K.: Google is rapidly turning into a formidable opponent to bff NVIDIA. https://www.techradar.com/pro/google-is-rapidly-turning-into-a-formidable-opponent-to-bff-nvidia-the-tpu-v5p-ai-chip-powering-its-hypercomputer-is-faster-and-has-more-memory-and-bandwidth-than-ever-before-beating-even-the-mighty-h100, Accessed 04 Aug 2024

81. Amin Vahdat, M.L.: Enabling next-generation AI workloads: announcing TPU v5p and AI hypercomputer. https://cloud.google.com/blog/products/ai-machine-learning/introducing-cloud-tpu-v5p-and-ai-hypercomputer, Accessed 04 Aug 2024
82. Velasco, A.: Google cloud unveils trillium, its 6th-gen TPU with a huge 4.7x AI performance leap. https://hothardware.com/news/google-cloud-unveils-trillium-its-6th-gen-tpu-with-a-47x-performance-leap, Accessed 04 Aug 2024
83. Vahdat, A.: Announcing trillium, the sixth generation of google cloud TPU. https://cloud.google.com/blog/products/compute/introducing-trillium-6th-gen-tpus, Accessed 04 Aug 2024
84. MICROSOFT: Microsoft's AI suppercomputer. https://news.microsoft.com/source/features/ai/openai-azure-supercomputer/, Accessed 04 Aug 2024
85. MICROSOFT: AI and the need for purpose-built cloud infrastructure. https://azure.microsoft.com/en-us/blog/ai-and-the-need-for-purposebuilt-cloud-infrastructure/, Accessed 04 Aug 2024
86. Ooko, S.O., Mukanyiligira, D., Munyampundu, J.P., Nsenga, J.: Edge AI-based respiratory disease recognition from exhaled breath signatures. In: 2021 IEEE Jordan International Joint Conference on Electrical Engineering and Information Technology (JEEIT), pp. 89–94 (2021)
87. Ooko, S.O., Mukanyiligira, D., Munyampundu, J.P., Nsenga, J.: Synthetic exhaled breath data-based edge AI model for the prediction of chronic obstructive pulmonary disease. In: 2021 International Conference on Computing and Communications Applications and Technologies (I3CAT), pp. 1–6 (2021)
88. Kumar, H.H., Karthik, V.R., Nair, M.K.: Federated k-means clustering: a novel edge AI based approach for privacy preservation. In: 2020 IEEE International Conference on Cloud Computing in Emerging Markets (CCEM), pp. 52–56 (2020)
89. Zhang, J., et al.: Anticoncealer: reliable detection of adversary concealed behaviors in edgeai-assisted IoT. IEEE Internet Things J. **9**(22), 22184–22193 (2022)
90. Wang, X., Han, Y., Leung, V.C.M., Niyato, D., Yan, X., Chen, X.: Convergence of edge computing and deep learning: a comprehensive survey. IEEE Commun. Surv. Tutor. **22**(2), 869–904 (2020)
91. Ray, P.P.: A review on tinyml: state-of-the-art and prospects. J. King Saud Univ. Comput. Inf. Sci. **34**(4), 1595–1623 (2022)
92. Rocha, D., Rocha, P., Ribeiro, J., Lopes, S.: Identification and classification of human body parts for contactless screening systems: an edge-AI approach, pp. 92–103 (06 2022)
93. coral: https://coral.ai/products/dev-board/
94. NVIDIA: https://www.nvidia.com/en-us/autonomous-machines/embedded-systems/jetson-nano/product-development/
95. QUALCOMM: https://www.qualcomm.com/products/mobile/snapdragon/laptops-and-tablets/snapdragon-x-plus
96. BEAGLEBOARD: https://www.beagleboard.org/boards/beaglebone-ai
97. INTEL: https://www.intel.vn/content/www/vn/vi/products/sku/240961/intel-core-ultra-9-processor-288v-12m-cache-up-to-5-10-ghz/specifications.html
98. HALIO: https://hailo.ai/products/ai-accelerators/hailo-8-ai-accelerator/
99. Tan, T., Cao, G.: Efficient execution of deep neural networks on mobile devices with NPU, pp. 283–298 (2021)

100. Apple: https://machinelearning.apple.com/research/neural-engine-transformers (2022)
101. clamchowder: https://chipsandcheese.com/2023/10/04/qualcomms-hexagon-dsp-and-now-npu/ (2023)
102. https://www.qualcomm.com/products/mobile/snapdragon/laptops-and-tablets/snapdragon-x-plus (2024)
103. NXP: https://www.nxp.com/products/processors-and-microcontrollers/arm-processors/i-mx-applications-processors/i-mx-8-applications-processors/coral-dev-board-tpu:CORAL-EDGE-TPU
104. Reidy, B.C., Mohammadi, M., Elbtity, M.E., Zand, R.: Efficient deployment of transformer models on edge TPU accelerators: a real system evaluation. In: Architecture and System Support for Transformer Models (ASSYST @ISCA 2023) (2023)
105. Hadnagy, Ã., FehÃ©r, B., KovÃcshÃzy, T.: Efficient implementation of convolutional neural networks on fpga. In: 2018 19th International Carpathian Control Conference (ICCC), pp. 359–364 (2018)
106. Yazici, M.T., Basurra, S., Gaber, M.M.: Edge machine learning: enabling smart internet of things applications. Big Data Cogn. Comput. 2(3) (2018)
107. NVIDIA: https://developer.nvidia.com/embedded/jetson-nano
108. Ray, P.P.: A review on tinyml: state-of-the-art and prospects. J. King Saud Univ. Comput. Inf. Sci. 34(4), 1595–1623 (2022)
109. Research, G.V.: Edge AI market size, share & trends analysis report by component (hardware, software, edge cloud infrastructure, services), by end-use industry, by region, and segment forecasts, 2023 - 2030. https://www.grandviewresearch.com/industry-analysis/edge-ai-market-report, Accessed 07 Aug 2024
110. Insights, F.B.: Cloud AI market size, share & Covid-19 impact analysis, by component (solution and services), by technology (machine learning (ml), deep learning, and natural language processing (NLP) and others), by function (finance, marketing & sales, supply chain management, human resources, and others), by end-users (BFSI, it and telecommunications, healthcare, retail and consumer goods, media and entertainment, and others), and regional forecast, 2023 – 2030. https://www.fortunebusinessinsights.com/cloud-ai-market-108878, Accessed 07 Aug 2024
111. viso.ai: Object counting. https://viso.ai/application/object-counting/, Accessed 05 Aug 2024
112. Qi, L., Li, J., N.A.H.W.: Security and privacy issues for artificial intelligence in edge-cloud computing. https://www.springeropen.com/collections/securityprivacyaiedgecloud, Accessed 07 Aug 2024
113. Report, I.: 2.9 billion people still offline. https://www.itu.int/en/mediacentre/Pages/PR-2021-11-29-FactsFigures.aspx, Accessed 05 Aug 2024
114. Sipola, T., Alatalo, J., Kokkonen, T., Rantonen, M.: Artificial intelligence in the IoT era: a review of edge AI hardware and software. In: 2022 31st Conference of Open Innovations Association (FRUCT), pp. 320–331 (2022)
115. Singh, R., Armour, S., Khan, A., Sooriyabandara, M., Oikonomou, G.: The advantage of computation offloading in multi-access edge computing. In: 2019 Fourth International Conference on Fog and Mobile Edge Computing (FMEC), pp. 289–294 (2019)
116. Tuvey, E.: What companies can do about cloud spend wastage. https://www.forbes.com/sites/richardnieva/2024/08/05/groq-funding-series-d-nvidia/?, Accessed 05 Aug 2024

117. Himeur, Y., Alsalemi, A., Bensaali, F., Amira, A., Al-Kababji, A.: Recent trends of smart nonintrusive load monitoring in buildings: a review, open challenges, and future directions. Int. J. Intell. Syst. **37**(10), 7124–7179 (2022)
118. Raychaudhuri, A., Mukherjee, A., De, D., Gill, S.S.: Green internet of things using mobile cloud computing: architecture, applications, and future directions. In: Green Mobile Cloud Computing, pp. 213–229. Springer (2022)
119. Schuster, R., Ramchandran, P.: Open edge computing: from vision to reality. In: Conference Proceedings–24-25 June 2016, p. 135 (2016)

A Novel Pseudoconvex Programming Approach for Multi-criteria Fuzzy Portfolio Optimization with Uncertainty Ratio

Nguyen Khac Trung[1], Nguyen Thi Xuan Hoa[2,3], Thai Minh Hanh[2,3],
Cu Xuan Quoc[2], Le Viet Bach[1], and Tran Ngoc Thang[1(✉)]

[1] Faculty of Mathematics and Informatics, Hanoi University of Science and
Technology, Hanoi, Vietnam
`thang.tranngoc@hust.edu.vn`
[2] School of Economics and Management, Hanoi University of Science and
Technology, Hanoi, Vietnam
[3] Center for Digital Technology and Economy (BK Fintech),
Hanoi University of Science and Technology, Hanoi, Vietnam

Abstract. Portfolio optimization remains a critical task in finance, aiming to attain an ideal equilibrium between the level of risk and the corresponding return. Many approaches often rely on metrics such as the Sharpe and Value at Risk (VaR) ratios to evaluate portfolio performance and risk exposure. This paper applies the above indicators with fuzzy numbers to incorporate uncertain factors into the model, thereby helping the model reflect reality more closely. Using these metrics in portfolio selection models poses challenges due to their objective functions' nonlinear and non-convex nature. Instead of using the existing heuristic approach, we propose a new approach based on pseudo-convex programming. Theoretically, we prove the pseudo-convexity of the fuzzified objective functions, thereby showing that the equivalent problem is pseudoconvex. Taking advantage of the elegant properties of pseudo-convex programming, we propose an efficient algorithm to tackle the multi-objective fuzzy portfolio optimization problem. Experimentally, we demonstrate the effectiveness of our approach through empirical analyses using real-world financial data and its superiority over multi-objective evolutionary algorithms such as NSGA-III and AGE-MOEA-II.

Keywords: Multi-criteria Pseudoconvex Programming · Value at Risk Ratio · Sharpe Ratio · Fuzzy Portfolio Selection

1 Introduction

Portfolio selection is a critical decision-making process in investment management that is intended to improve capital allocation across various asset classes

This work was funded by Rikkeisoft Corporation and supported by Center for Digital Technology and Economy (BK Fintech), Hanoi University of Science and Technology.

S. Thomassey et al. (Eds.): RAIDS 2024, LNICST 673, pp. 218–231, 2026.
https://doi.org/10.1007/978-3-032-14055-5_15

to construct an optimal portfolio. The primary goals of portfolio selection are to optimize the anticipated returns while mitigating the associated risks in accordance with the risk tolerance and the investor's investment horizon.

Since the establishment of the Mean-Variance model, also known as the Modern Portfolio Theory (MPT), by Harry Markowitz, the concept of portfolio selection gained prominence. Markowitz's model quantifies risk as the variance or standard deviation of portfolio returns and demonstrates that investors can reduce overall portfolio risk through diversification across different assets with imperfect correlations. There have been other proposed extensions and alternative methods after Markowitz's contribution, enhancing portfolio selection models and integrating additional factors such as risk aversion index [5] or entropy [20]. This research delves deeper into the analysis of the Sharpe ratio and the Value at Risk (VaR) ratio. By incorporating both measures, investors and portfolio managers can obtain a more comprehensive understanding of the risk-return tradeoff. The Sharpe ratio helps evaluate the overall risk-adjusted performance. In contrast, the VaR ratio complements this by explicitly accounting for the potential adverse risk, which is a crucial consideration in risk management and capital allocation decisions.

Due to the assumption of accurate and deterministic inputs, traditional portfolio selection methods may struggle in complicated financial markets. Financial data's unpredictability, volatility, and asset correlations are a significant issue. Fuzzy numbers offer a solution by providing a flexible and intuitive framework for modeling uncertainty and imprecision. In light of this, several distinct portfolio selection models incorporating fuzziness have been constructed using a variety of methods, such as enhancing decision processes [6,11] or utilizing set theory [19]. Hybrid fuzzy models, like neural networks [2], stochastic models [9], and genetic algorithms [14] also offer significant advantages.

One of the significant fuzzy portfolio selection models was developed by Kar et al. in 2019 (see [7]). However, the characteristics of the goal functions were not exploited. Thus, the current strategy employed is still based on heuristic methods such as multiobjective evolutionary algorithms. This strategy suffers from a significant drawback, as it requires a substantial amount of computation for a good enough approximation of the Pareto front. By uncovering the interesting characteristics of the fuzzy portfolio selection problem, we consider it a multi-objective pseudoconvex programming problem. To solve the main problem, we transformed it into a single-objective pseudoconvex programming problem. Some efficient algorithms can find their global optimal solution, such as gradient methods in [15] and references therein. We have proved that the objective functions remain pseudo-convex for any form of membership function that is used in the model. It allows the model to be adaptable to various types of practical situations. Computational experiments show that the algorithm works accurately and efficiently for practical datasets. The source code can be packaged into an installable library in Python or MATLAB for convenient development. Similar to the convex properties of the Markowitz model, the pseudoconvex properties of the fuzzy portfolio optimization model can be utilized to solve related classes of problems, such as the optimization problem over the Pareto frontier.

The following section presents the primary methodology of portfolio selection. Section 3 discusses the pseudoconvex programming problem, outlining our methods and supplying the essential definitions. In Sect. 4, we demonstrate our methodology's reliability by applying it to real-world portfolio selection challenges. In Sect. 5, summarize the content of the paper.

2   Multi-criteria Portfolio Selection Problem

To guarantee that all available funds are entirely invested without exceeding or falling below the total investment amount in real-world investment situations, a constraint is in place as the following equation:

$$\sum_{i=1}^{n} x_i = 1; \quad x_i \geq 0, i = \overline{1,n}, \tag{1}$$

where $x = (x_1, ..., x_n)$ is denoted as the portfolio vector, which represents the allocation of proportion x_i into the ith asset. The anticipated average return from investing in a combination of n assets is represented by the expected return of a portfolio, denoted by $E(x)$. Specifically, the expected return is determined by multiplying the expected return of each asset $\mathcal{L}_i$ by the respective allocation weight x_i and summing the products across all assets. It can be denoted as

$$E(x) = \sum_{i=1}^{n} \mathcal{L}_i x_i. \tag{2}$$

The variance of returns, or portfolio risk, assesses the volatility of a portfolio of n assets, denoted by $V(x)$. The variance of returns is calculated by multiplying each pair of asset covariances σ_{ij} by the corresponding allocation weights (x_i and x_j) and summing these products across all asset pairs as follows:

$$V(x) = \sum_{i=1}^{n} \sum_{j=1}^{n} \sigma_{ij} x_i x_j. \tag{3}$$

The Markowitz model for multi-criteria optimization optimizes two conflicting goals, such as return and risk, presented by

$$\begin{aligned} &\text{Max } E(x) \\ &\text{Min } V(x) \\ &\text{s.t. } \sum_{i=1}^{n} x_i = 1; \quad x_i \geq 0, i = \overline{1,n}. \end{aligned} \tag{4}$$

The Sharpe Ratio is a measure of the risk-adjusted return of an investment. It is calculated by deducting the risk-free rate r_f from the expected return of a portfolio and then dividing the result by the standard deviation of the portfolio's returns, that is

$$\mathrm{SR}(x) = \frac{E(x) - r_f}{\sqrt{V(x)}}. \tag{5}$$

Value at Risk (VaR) is an additional statistical measure frequently employed as a risk management tool to evaluate and monitor the potential downside risk of investment portfolios. For a random variable x representing the portfolio's return or loss, at a level of confidence β, the VaR is described as

$$\mathrm{VaR}_\beta(x) = \inf\{x \in \mathbb{R} | \mathbb{P}(X \leq x) \geq 1 - \beta\}. \tag{6}$$

Our research integrates both the Sharpe Ratio and Value at Risk (VaR) ratio using fuzzy numbers. This helps us create a comprehensive risk management framework that considers both the risk-adjusted return and the possible adverse risk of an investment portfolio in a more sophisticated and adaptable way.

3   Multi-criteria Fuzzy Portfolio Selection

3.1   Fuzzy Number

Take into account the subsequent mapping: $\mu_{\mathcal{F}} : \mathbb{X} \to [0,1]$ with a universal set $\mathbb{X}$. This is referred to as the membership mapping. Therefore, the fuzzy subset is described

Definition 1. *(see in [7]) A fuzzy subset $\mathcal{F}$ is defined as*

$$\mathcal{F} = \{(x, \mu_{\mathcal{F}}(x)) : x \in \mathbb{X}\}. \tag{7}$$

A set that represents all elements in a fuzzy number that have a membership degree larger than or equal to a specific threshold is referred to as a level set. It offers a method for representing fuzzy numbers by dividing them into a sequence of nested intervals. We considered the following definition.

Definition 2. *The γ-level set of $\mathcal{F}$, represented by $[\mathcal{F}]^\gamma$, is described as*

$$[\mathcal{F}]^\gamma = \{\gamma \in [0,1], x \in \mathbb{X} : \mu_{\mathcal{F}}(x) \geq \gamma\}. \tag{8}$$

Using the level set, the following definitions and theorems are established (see in [7]). Two fuzzy numbers, $\mathcal{A}$ and $\mathcal{B}$, are identified with the corresponding γ-level set $[\mathcal{A}]^\gamma = [a_1(\gamma), a_2(\gamma)]$ and $[\mathcal{B}]^\gamma = [b_1(\gamma), b_2(\gamma)]$ respectively. We can define the statistical measures by utilizing their level set as follows.

Definition 3. *The mean and variance of $\mathcal{A}$ is*

$$E[\mathcal{A}] = \int_0^1 \gamma \{a_1(\gamma) + a_2(\gamma)\}\, \mathrm{d}\gamma, \tag{9}$$

$$V[\mathcal{A}] = \frac{1}{2} \int_0^1 \gamma \{a_2(\gamma) - a_1(\gamma)\}^2\, \mathrm{d}\gamma. \tag{10}$$

Definition 4. *The relationship between $\mathcal{A}$ and $\mathcal{B}$ can be described using the covariance*

$$Cov\left(\mathcal{A}, \mathcal{B}\right) = \frac{1}{2} \int_0^1 \gamma \left\{a_2\left(\gamma\right) - a_1\left(\gamma\right)\right\} \left\{b_2\left(\gamma\right) - b_1\left(\gamma\right)\right\} \mathrm{d}\gamma. \tag{11}$$

Using probability formulae, the expected value and variance of a weighted combination of the two fuzzy numbers can be computed. Presented here is the theorem considering two real numbers $\lambda_1, \lambda_2 \in \mathbb{R}$.

Theorem 1. *The following equations give the mean value and variance of weighted sums of two fuzzy numbers:*

$$E\left[\lambda_1 \mathcal{A} + \lambda_2 \mathcal{B}\right] = \lambda_1 E\left[\mathcal{A}\right] + \lambda_2 E\left[\mathcal{B}\right], \tag{12}$$

$$V\left[\lambda_1 \mathcal{A} + \lambda_2 \mathcal{B}\right] = \lambda_1^2 V\left[\mathcal{A}\right] + \lambda_2^2 V\left[\mathcal{B}\right] + 2\left|\lambda_1 \lambda_2\right| Cov\left(\mathcal{A}, \mathcal{B}\right). \tag{13}$$

3.2   Value at Risk

Credibility [12] can be defined as a mapping from the power set Λ of $\Theta \neq \emptyset$ to the interval $[0, 1]$. To ensure mathematical properties, this measure meets four axioms: Normality, Monotonicity, Duality, and Maximality. Based on these axioms, we construct Value at Risk (VaR) for a fuzzy variable $\mathcal{A}$, representing investment loss, at a confidence level $\beta \in (0, 1)$.

Definition 5. *The VaR of $\mathcal{A}$ is the infimum of r such that the credibility of the event $\mathcal{A} \leq r$ is at least β*

$$\mathrm{VaR}[\mathcal{A}] = \inf\{r | \mathrm{Credibility}\{\mathcal{A} \leq r\} \geq \beta\}. \tag{14}$$

The linearity property of the VaR function, which is established in Theorem 2 and Theorem 3, was also demonstrated by Li [8]. This property is crucial for proving the pseudoconvexity of the main problem.

Theorem 2. *Suppose that $\mathcal{A}$ and $\mathcal{B}$ are mutually independent fuzzy variables. Then, for any $\beta \in (0, 1]$, we have*

$$\mathrm{VaR}\left[\mathcal{A} + \mathcal{B}\right] = \mathrm{VaR}\left[\mathcal{A}\right] + \mathrm{VaR}[\mathcal{B}] \tag{15}$$

Theorem 3. *Suppose that $\mathcal{A}$ is a fuzzy variable. Then, for any real number $\alpha \geq 0$, we have*

$$\mathrm{VaR}\left[\alpha \mathcal{A}\right] = \alpha \mathrm{VaR}\left[\mathcal{A}\right] \tag{16}$$

With different purposes and applications in modeling uncertainty and imprecision, many types of fuzzy numbers provide more flexibility in data representation. For example, triangular fuzzy numbers are defined by a simple, piecewise linear membership function forming a triangular shape, while Gaussian fuzzy numbers are represented by a smooth, bell-shaped membership function.

The following remarks present fundamental formulas for calculating essential statistical properties of different forms of fuzzy numbers, notably triangular fuzzy numbers and Gaussian fuzzy numbers, and will be used in the experiment.

Remark 1. (see [7,18]) If $\mathcal{A} = (\alpha_1, \alpha_2, \alpha_3)$ is a triangular fuzzy number (TFN), then

$$E\left[\mathcal{A}\right] = \frac{\alpha_1 + 4\alpha_2 + \alpha_3}{6}, \tag{17}$$

$$V\left[\mathcal{A}\right] = \frac{(\alpha_2 - \alpha_1)^2 + (\alpha_3 - \alpha_2)^2 + (\alpha_2 - \alpha_1)(\alpha_3 - \alpha_2)}{18}, \tag{18}$$

$$\text{VaR}\left[\mathcal{A}\right] = \begin{cases} 2\beta\alpha_2 + (1 - -2\beta)\alpha_1, & \text{if} \quad \beta \leq 0.5, \\ (2\beta - 1)\alpha_3 + (2 - -2\beta)\alpha_2, & \text{if} \quad \beta > 0.5. \end{cases} \tag{19}$$

Remark 2. (see [8]) If $\mathcal{A} = N(e, \sigma)$ is a Gaussian (normal) fuzzy number, then

$$E\left[\mathcal{A}\right] = e, \tag{20}$$

$$V\left[\mathcal{A}\right] = \sigma^2, \tag{21}$$

$$\text{VaR}\left[\mathcal{A}\right] = e - (\ln(1 - \beta) - \ln\beta)\sqrt{6}\sigma/\pi \tag{22}$$

Our experiments will demonstrate that the wide range of fuzzy numbers can be effectively utilized in our optimization problem, regardless of the linearity of the membership function. This highlights the flexibility and adaptability of our approach, showing that it can handle various types of fuzzy number representations, whether linear or nonlinear. By leveraging this variety, our method is versatile and robust in addressing different scenarios and challenges within the optimization process.

3.3   Integrating Sharpe and VaR Ratios with Fuzzy Number

The Sharpe Ratio is a financial metric used to evaluate the risk-adjusted return of an investment or portfolio. The ratio measures how much excess return can be received for the extra volatility endured for holding a riskier asset. A higher Sharpe Ratio indicates a more attractive risk-adjusted return. It is calculated as follows:

Definition 6. *Let a fuzzy number $\mathcal{A}$ represent the portfolio return; Sharpe ratio can be defined as*

$$\text{SR}\left[\mathcal{A}\right] = \frac{E\left[\mathcal{A}\right] - R_b}{V\left[\mathcal{A}\right]}, \tag{23}$$

where the variable R_b represents the return of a standard portfolio, which is typically a market index; $E\left[\mathcal{A}\right]$ and $V\left[\mathcal{A}\right]$ are mean and variance of the portfolio return.

The Value at Risk ratio (VR) is displayed as a systematic risk-based ratio and is utilized to determine the level of risk associated with investments. The greater the value, the more favorable the investment opportunities. It can be defined as follows:

Definition 7. *The investment's return is indicated as a fuzzy number $\mathcal{A}$. Subsequently, for any $\beta \in (0,1)$, its β-Value at Risk ratio is defined as*

$$\mathrm{VR}\left[\mathcal{A}\right] = \frac{E\left[\mathcal{A}\right] - R_b}{\mathrm{VaR}\left[\mathcal{A}\right]}. \tag{24}$$

Maximizing both the Sharpe ratio and the VaR ratio provides a well-rounded approach to optimizing investment portfolios by enhancing risk-adjusted returns, improving portfolio stability, aiding in informed decision-making, and boosting investor confidence.

3.4   Problem Formulation

Consider a financial market where investors distribute their funds across n risky assets using proportions x_i, with $i = 1, ..., n$. Each asset is characterized by its fuzzy return $\mathcal{A}_i$. Then, we have the following bi-criteria optimization problem:

$$\mathrm{Max}\left(\mathrm{SR}\left[\sum_{i=1}^{n}\mathcal{A}_i x_i\right], \mathrm{VR}\left[\sum_{i=1}^{n}\mathcal{A}_i x_i\right]\right)^{T} \tag{FBO}$$
$$\mathrm{s.t.} \sum_{i=1}^{n} x_i = 1; \quad x_i \geq 0, i = \overline{1, n}.$$

A portfolio's performance is evaluated using the suggested model, which takes into account both systematic and nonsystematic risks. The optimal solutions are obtained in order to optimize both the standard deviation and the volatility.

3.5   Transformation to Deterministic Model

This work aims to address Problem (FBO) by showcasing its favorable attribute derived from the properties of the objective functions.

Definition 8. (Pseudoconvex Function (see [10])). *$f : \mathbb{R}^n \to \mathbb{R}$ is called pseudoconvex on a convex set $X \subseteq \mathbb{R}^n$ if $\forall x, y \in X$, the following inequality is valid*

$$f(y) < f(x) \Rightarrow \langle \nabla f(x), y - x \rangle < 0. \tag{25}$$

If the function f is pseudoconvex, then its negation $-f$ is termed a pseudoconcave function.

We will now verify the pseudoconcavity of the ratio defined in Eqs. (23) and (24).

Proposition 1. *Suppose that the fuzzy returns of all securities are independent of each other, $\mathrm{SR}\left[\mathcal{A}\right]$ and $\mathrm{VR}\left[\mathcal{A}\right]$ are pseudoconcave functions.*

Proof. Due to Definition 3 and the fact that R_b is a constant, $E\left[\mathcal{A}\right] - R_b$ is positive linear. As defined above, assume $\mathcal{A}$ and $\mathcal{B}$ are fuzzy numbers with γ-level sets. Derived from the AM-GM inequality, we have

$$2Cov\left(\mathcal{A}, \mathcal{B}\right) \leq V\left[\mathcal{A}\right] + V\left[\mathcal{B}\right].$$

Multiply both sides of the inequality with $\lambda(1 - \lambda)$, with $\lambda \in (0, 1)$:

$$2\lambda(1 - \lambda)Cov\left(\mathcal{A}, \mathcal{B}\right) \leq \lambda(1 - \lambda)V\left[\mathcal{A}\right] + \lambda(1 - \lambda)V\left[\mathcal{B}\right]$$

$$\Leftrightarrow (1 - \lambda)^2 V\left[\mathcal{B}\right] + \lambda^2 V\left[\mathcal{A}\right] + 2\lambda(1 - \lambda)Cov\left(\mathcal{A}, \mathcal{B}\right) \leq (1 - \lambda)V\left[\mathcal{B}\right] + \lambda V\left[\mathcal{A}\right]$$

$$\Leftrightarrow V\left[(1 - \lambda)\mathcal{B} + \lambda\mathcal{A}\right] \leq (1 - \lambda)V\left[\mathcal{B}\right] + \lambda V\left[\mathcal{A}\right]. \text{ (According to Theorem 1)}$$

Therefore, $V\left[\mathcal{A}\right]$ is a convex function. Furthermore, due to the independence of the returns of all securities, the Value at Risk (VaR$[\mathcal{A}]$) is a linear function.

Given two differentiable functions f_1 and f_2 on a set X, where f_1 is positive and concave while f_2 is positive and convex, the fractional function $\frac{f_1}{f_2}$ is pseudo-concave on X [1]. As a result, SR$\left[\mathcal{A}\right]$ and VR$\left[\mathcal{A}\right]$ are pseudoconcave functions, leading to SR$^*\left[\mathcal{A}\right] = -SR\left[\mathcal{A}\right]$ and VR$^*\left[\mathcal{A}\right] = -VR\left[\mathcal{A}\right]$ being pseudoconvex.

Based on Proposition 1, we have demonstrated that Problem (FBO) by taking the negative of both the objective function required, the problem becomes a multiobjective pseudoconvex function. The problem can be expressed in the following manner

$$\text{Min}\left(\text{SR}^*\left[\sum_{i=1}^{n} \mathcal{A}_i x_i\right], \text{VR}^*\left[\sum_{i=1}^{n} \mathcal{A}_i x_i\right]\right)^T \tag{FBOP}$$

$$\text{s.t.} \sum_{i=1}^{n} x_i = 1; x_i \geq 0, i = \overline{1, n}.$$

To save considerable computational expense, we propose converting Problem (FBOP) into the following problem, as inspired by [17], where $y = (y_1, y_2)$, $v = (v_1, v_2)$ are the parameters

$$\text{Min}\left\{\text{Max}\left\{\frac{\text{SR}^*[\sum_{i=1}^{n} \mathcal{A}_i x_i] - y_1}{v_1}, \frac{\text{VR}^*[\sum_{i=1}^{n} \mathcal{A}_i x_i] - y_2}{v_2}\right\}\right\} \tag{BFBOP}$$

$$\text{s.t.} \sum_{i=1}^{n} x_i = 1; x_i \geq 0, i = \overline{1, n}.$$

This is a pseudoconvex programming problem, in which the local optimal solution is also the global one. We address Problem (BFBOP) by performing optimization on the branch defined by projecting $y = (y_1, y_2)$ with respect to the vector $v = (v_1, v_2)$. The set of nondominated solutions collectively forms a reference front, approximately to the Pareto front - the set of optimum solutions. This reference front is limited by two endpoints, which are the solutions obtained when independently optimizing each objective function in Problem (FBOP). We called the two problems (FO-SR) for optimizing the Sharpe ratio and (FO-VR) for optimizing the VaR ratio. The outcomes of the experiments below evidence the practicality of this approach.

4   Computational Experiments

The experiments conducted in this paper are implemented in Python programming language using Google Colab with Intel(R) Xeon(R) CPU @ 2.20GHz CPU. We provide results showing the usability and adaptability of our approach. We also compare the approach to different evolutionary algorithms, showing the advantage of deterministic methods over heuristic methods in some cases.

4.1   Test Case 1

Take into account 20 different securities with the corresponding forecast price during the first four months of 2016 given in Table 1 (see [7]). The forecast prices have been transformed into triangular fuzzy numbers from real-life data through the granular computing method [21].

Table 1. Returns of 20 securities in the form of triangular fuzzy numbers.

Security code	Return	Security code	Return
000001.SZ	$(0.8803, 0.9923, 1.1083)$	000016.SZ	$(0.4476, 0.9854, 1.2594)$
000002.SZ	$(0.8796, 1.0018, 1.0838)$	000017.SZ	$(0.6494, 0.9999, 1.3699)$
000004.SZ	$(0.8408, 1.0150, 1.3210)$	000018.SZ	$(0.7589, 1.0089, 1.3729)$
000005.SZ	$(0.7130, 1.0324, 1.6114)$	000019.SZ	$(0.7176, 1.0150, 1.2800)$
000006.SZ	$(0.7548, 1.0110, 1.3230)$	000020.SZ	$(0.7585, 1.0268, 1.4698)$
000009.SZ	$(0.7637, 1.0053, 1.2073)$	000021.SZ	$(0.8993, 1.0100, 1.3110)$
000010.SZ	$(0.7686, 1.0025, 1.4505)$	000022.SZ	$(0.8251, 0.9878, 1.0248)$
000011.SZ	$(0.7894, 1.0141, 1.4221)$	000023.SZ	$(0.8114, 1.0044, 1.2644)$
000012.SZ	$(0.8446, 0.9888, 1.1218)$	000024.SZ	$(0.8529, 1.0335, 1.2825)$
000014.SZ	$(0.7490, 1.0127, 1.2777)$	000025.SZ	$(0.7177, 1.0868, 1.5928)$

By specifying the confidence threshold for the VaR at 0.9, we are able to compute the mean value (or expected return), variance, and VaR for each triangular fuzzy number. This computation is carried out using the methodologies outlined in Remark 1 in conjunction with Eqs. (20), (21), and (22). Table 2 presents the calculated values for these parameters corresponding to the first five security codes.

Table 2. Expected Return, Variance, and Value at Risk of the first five securities.

Security code	000001.SZ	000002.SZ	000004.SZ	000005.SZ	000006.SZ
$E(\mathcal{A})$	0.9930	0.9951	1.0370	1.0757	1.0203
$V(\mathcal{A})$	0.0022	0.0018	0.0098	0.0346	0.0135
$\mathrm{VaR}(\mathcal{A})$	1.0851	1.06741	1.2598	1.4956	1.2606

To determine the reference front of the Problem (FBOP), the risk-free rate (R_b) for both targets is fixed at 0.05. We first solve Problem (FO-SR) and (FO-VR) respectively to indicate the two endpoints, which are denoted as x^{SR} and x^{VR}. The outcomes are outlined in Table 3.

Table 3. Sharpe ratio and Value at Risk ratio of each solution.

Problem	Solution	SR	VR
(FO-SR)	x^{SR}	537.0345	0.8854
(FO-VR)	x^{VR}	487.3954	0.9012

After identifying the two endpoints, we proceed to address Problem (BFBOP) by employing a projection method originating from points $y = (0,0)$, relative to the vector $v = (v_1, v_2)$, situated within the domain delineated by the branch derived from y, individually for each x^{SR} and x^{VR}.

Figure 1 illustrates the Pareto-optimal value set obtained for the bi-criteria fuzzy portfolio optimization problem (BFBOP) after solving it using the proposed pseudoconvex programming approach. The plot demonstrates the trade-off between the two opposing objectives. The findings depicted within the image illustrate the efficacy of the proposed approach in generating a relatively smooth reference front with these solutions.

Our deterministic approach provides consistent and repeatable results, reducing the variability and randomness inherent in heuristic methods even with state-of-the-art multi-objective evolutionary algorithms. This is particularly advantageous in applications where minor deviations, such as financial modeling, engineering design, and critical system optimizations, can lead to significant impacts.

Hypervolume [4] is a crucial metric in multi-objective optimization used to assess the quality of a set of solutions by measuring the volume of the objective space they dominate relative to a reference point. It quantifies how well a set of solutions covers the objective space, reflecting the tradeoffs among different objectives. A larger hypervolume indicates that the solutions span a more extensive area of the objective space, suggesting a better approximation of the Pareto front and a more diverse set of solutions. This measure is vital for comparing the performance of different optimization algorithms, as it highlights the extent to which the solutions offer balanced and comprehensive tradeoffs between multiple

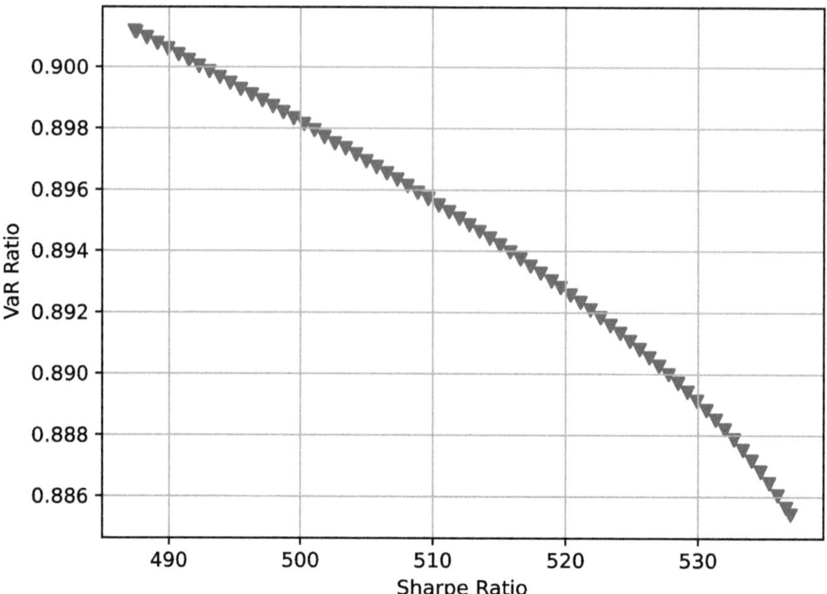

Fig. 1. Pareto-optimal value set for test case 1.

objectives. We can use Hypervolume as a metric to compare different evolutionary algorithms with our approach.

Table 4 gives us insights into the advantage our approach has over different evolutionary algorithms such as AGE-MOEA-II [13] and NSGA-III [3]. The time to reach the Hypervolume expected in other heuristic algorithms is significantly longer than our deterministic approach. Also, in the same time interval of executing different algorithms, our projection on a branch gives a much more accurate solution, closer to the desired Pareto front, than both evolutionary algorithms. These results help us understand the superiority of the deterministic approach over different heuristic algorithms.

4.2   Test Case 2

We simulated a new test case to show that our optimization model is not limited by the linearity of the fuzzy number we utilized. Except for the traditional triangular and trapezoidal fuzzy numbers we usually use, other nonlinear forms can also produce a solid solution by using our approach.

We generate 20 securities similar to test case 1 by randomizing spot prices p_i between 50 and 150, the estimated closing prices p'_i within +/- 10 of the spot prices, and the estimated dividends d_i between 0 and 5. We calculate the returns as fuzzy numbers by using the following equation:

$$\mathcal{A}_i = \frac{p'_i + d_i - p_i}{p_i} \tag{26}$$

Table 4. Comparison of different algorithms by time to find the nondominated solutions and Hypervolume.

Algorithm	Parameters		Hypervolume	Time
NSGA-III	population_size	max_generations		
	75	150	2.390428	6.042 s
	100	500	5.296685	20.284 s
AGE-MOEA-II	population_size	max_generations		
	50	225	3.901911	6.045 s
	100	250	5.297609	14.251 s
Projectiles (Ours)	**800 branches**		**5.299346**	**6.069 s**

We then transform the fuzzy numbers generated into Gaussian (normal) fuzzy numbers with mean as the returns calculated using the equation above and standard deviation by assuming a fixed percentage deviation for variance is 10%. Utilizing the appropriate measures provided by Li [8] in Remark 2, we optimize Problem (BFBOP) and produce the nondominated solutions, which are corresponding to the Pareto-optimal value set depicted in Fig. 2.

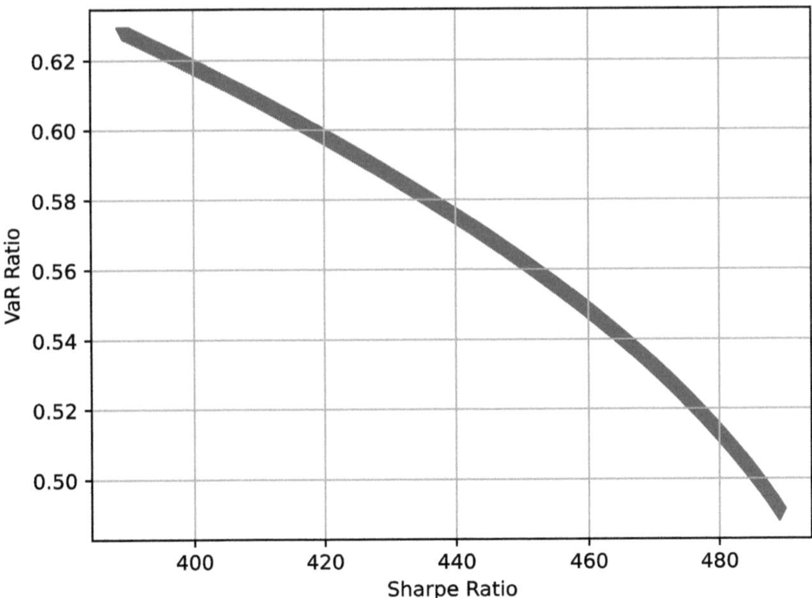

Fig. 2. Pareto-optimal value set for test case 2.

The results of our approach in both Test Case 1 and Test Case 2 demonstrate its versatility and adaptability in handling any fuzzy number, provided that the appropriate measures are applied. This capability is not limited by the shape or linearity of the fuzzy number's membership function, making it highly flexible for various modeling needs. Specifically, both triangular and Gaussian fuzzy numbers—two common representations in fuzzy set theory—yield accurate and well-defined Pareto-optimal fronts. These fronts represent a set of optimal solutions where no objective can be improved without compromising another, which is crucial in multi-objective optimization problems.

By generating Pareto-optimal fronts, our method equips investors with robust decision-making tools, allowing them to derive effective strategies considering risk and return trade-offs. Moreover, this flexibility supports comprehensive analysis in portfolio markets, offering investors insightful perspectives on market dynamics and uncertainty under different fuzzy conditions. Thus, the approach ensures that investors are better positioned to navigate the complexities of the market while optimizing their portfolios in uncertain environments.

5   Conclusion

This paper proposes a new method for solving the multi-criteria fuzzy portfolio optimization problem with the Sharpe and Value-at-Risk ratios. The model can effectively handle uncertainties and imprecise data inherent in practical investment scenarios by incorporating fuzzy numbers. We then formulate the problem as a pseudoconvex programming model by exploiting the pseudoconcavity property of the fuzzified Sharpe ratio and VaR ratio functions. Computational experiments on real-world financial data demonstrated the effectiveness of our approach in generating a reference front that approximates the actual Pareto optimal set. The reference front provides various non-dominated solutions, allowing investors to make informed decisions based on their specific risk-return preferences. The experiments also show many advantages of our deterministic approach over the recent heuristics approach, as well as the flexibility of using fuzzified information in our optimization model. For future work, we aim to handle additional real-world constraints and criteria relevant to portfolio selection and explore efficient solution algorithms with neurodynamic approach [16] or different approaches for the pseudoconvex formulation.

References

1. Avriel, M., Diewert, W.E., Schaible, S., Zang, I.: Generalized concavity. SIAM (2010)
2. Das, R., Sen, S., Maulik, U.: A survey on fuzzy deep neural networks. ACM Comput. Surv. (CSUR) **53**(3), 1–25 (2020)
3. Deb, K., Jain, H.: An evolutionary many-objective optimization algorithm using reference-point-based nondominated sorting approach, part i: solving problems with box constraints. IEEE Trans. Evolutionary Comput. **18**(4), 577–601 (2013)
4. Guerreiro, A.P., Fonseca, C.M., Paquete, L.: The hypervolume indicator: computational problems and algorithms. ACM Comput. Surv. (CSUR) **54**(6), 1–42 (2021)

5. Hoang, D.M., Thang, T.N., Tu, N.D., Hoang, N.V.: Stochastic linear programming approach for portfolio optimization problem. In: 2021 IEEE International Conference on Machine Learning and Applied Network Technologies (ICMLANT), pp. 1–4. IEEE (2021)
6. Hsu, Y.L., Lee, C.H., Kreng, V.B.: The application of fuzzy delphi method and fuzzy ahp in lubricant regenerative technology selection. Expert Syst. Appl. **37**(1), 419–425 (2010)
7. Kar, M.B., Kar, S., Guo, S., Li, X., Majumder, S.: A new bi-objective fuzzy portfolio selection model and its solution through evolutionary algorithms. Soft. Comput. **23**, 4367–4381 (2019)
8. Li, X.: Credibilistic programming. Springer (2013)
9. Lwin, K., Qu, R.: A hybrid algorithm for constrained portfolio selection problems. Appl. Intell. **39**, 251–266 (2013)
10. Mishra, S.K., Upadhyay, B.B.: Pseudolinear functions and optimization. CRC Press (2014)
11. Narang, M., Joshi, M.C., Bisht, K., Pal, A.: Stock portfolio selection using a new decision-making approach based on the integration of fuzzy cocoso with heronian mean operator. Decision Mak. Appl. Manag. Eng. **5**(1), 90–112 (2022)
12. Pahade, J.K., Jha, M.: Credibilistic variance and skewness of trapezoidal fuzzy variable and mean-variance-skewness model for portfolio selection. Results Appl. Math. **11**, 100159 (2021)
13. Panichella, A.: An improved pareto front modeling algorithm for large-scale many-objective optimization. In: Proceedings of the Genetic and Evolutionary Computation Conference, pp. 565–573 (2022)
14. Saborido, R., Ruiz, A.B., Bermúdez, J.D., Vercher, E., Luque, M.: Evolutionary multi-objective optimization algorithms for fuzzy portfolio selection. Appl. Soft Comput. **39**, 48–63 (2016)
15. Thang, T.N., Hai, T.N.: Self-adaptive algorithms for quasiconvex programming and applications to machine learning. Comput. Appl. Math. **43**(4), 249 (2024)
16. Thang, T.N., Hoang, D.M., Dung, N.V.: A neurodynamic approach for a class of pseudoconvex semivectorial bilevel optimization problems. Optimization Methods Softw., 1–28 (2024)
17. Vuong, N.D., Thang, T.N.: Optimizing over pareto set of semistrictly quasiconcave vector maximization and application to stochastic portfolio selection. J. Industr. Manag. Optimization **19**, 1999–2019 (2023)
18. Wan, S.P., Li, D.F., Rui, Z.F.: Possibility mean, variance and covariance of triangular intuitionistic fuzzy numbers. J. Intell. Fuzzy Syst. **24**(4), 847–858 (2013)
19. Wang, X., Wang, B., Liu, S., Li, H., Wang, T., Watada, J.: Fuzzy portfolio selection based on three-way decision and cumulative prospect theory. Int. J. Mach. Learn. Cybern. **13**(1), 293–308 (2022)
20. Zhou, J., Li, X., Pedrycz, W.: Mean-semi-entropy models of fuzzy portfolio selection. IEEE Trans. Fuzzy Syst. **24**(6), 1627–1636 (2016)
21. Zhu, X., Pedrycz, W., Li, Z.: Granular models and granular outliers. IEEE Trans. Fuzzy Syst. **26**(6), 3835–3846 (2018)

Preserving User Privacy in Retrieval Augmented Generation: A Novel Approach Using Local Placeholder Tagging

Thang Nguyen Xuan[(✉)], Vinh Nguyen Thanh, Thuy Duong Nguyen Duy,
Son Tran Huy Hoang, Gia Bao Nguyen, and Thao Nguyen Thi Ngoc

Hanoi University, Trung Van Ward, Nam Tu Liem District, Hanoi, Vietnam
`nxthang@hanu.edu.vn`

Abstract. Retrieval Augmented Generation (RAG) is a popular approach that enhances the accuracy of Large Language Models (LLMs) by leveraging a knowledge base. It is rapidly becoming integral tools across various applications. However, as the use of RAG continues to expand, so do the challenges associated with their deployment, particularly in terms of data privacy. As a part of RAG pipeline, user query and all retrieved documents should be sent as a prompt to the LLM providers, leaving them open to privacy hazards such data leaks or illegal access. This study presents RLPT, a framework designed to enhance user privacy in RAG. It achieves this by identifying and eliminating sensitive information from user inputs before sending them to the LLM. The RLPT framework utilizes a local LLM to rapidly identify sensitive information in user input and subsequently replaces it with distinctive placeholders. These placeholders are used to indicate and hide the actual sensitive data, ensuring that the LLM does not capture the original sensitive information during prompt processing. The framework is evaluated using a dataset consisting of 4000 synthesized context documents. The results indicate that it is capable of accurately detecting and filtering privacy and sensitive information, achieving a high accuracy rate of 88,7%.

Keywords: Retrieval-augmented generation · Large Language Model · Privacy protections · Data anonymization

1 Introduction

Retrieval Augmented Generation (RAG) is an AI framework that improves the quality of responses generated by LLM. It does this by incorporating external sources of knowledge to enhance the LLM's internal representation of information. RAG empowers the LLM by providing access to up-to-date and brand-specific information, enabling it to produce high-quality responses. This technology bridges the gap between the static knowledge of traditional LLMs and has transformed the way AI systems engage with, comprehend, and generate human language. RAG plays a vital role in enhancing the adaptability and intelligence of language models, with applications ranging from advanced chatbots to complex content creation tools [1].

S. Thomassey et al. (Eds.): RAIDS 2024, LNICST 673, pp. 232–245, 2026.
https://doi.org/10.1007/978-3-032-14055-5_16

However, the widespread adoption of RAG has also raised significant privacy concerns, particularly regarding the protection of sensitive data during interactions with the LLM providers. RAG model operates through two distinct stages, namely retrieval and generation, as depicted in Fig. 1. Upon the entry of a user query, relevant knowledge is initially extracted from an external vector database. Subsequently, the retrieved data is combined with the original query to create a prompt to a LLM. Leveraging its pretrained knowledge and retrieved data, the LLM proceeds to generate an appropriate response [2].

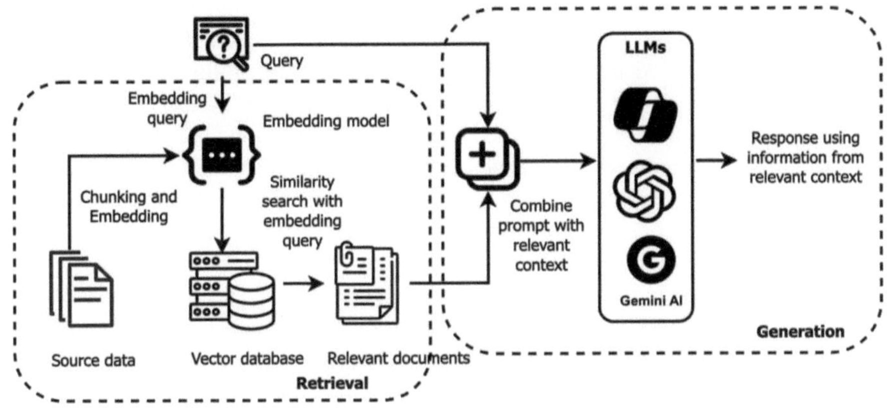

Fig. 1. Operation of RAG system

In many applications, the retrieval dataset can contain sensitive and valuable domain-specific information [3, 4]. For example, the relevant documents may contain Personally Identifiable Information (PII) [5] or confidential data like dates, product names, specifications, source code, or project briefs. Furthermore, the dataset may include patent knowledge for engineering design [6] or patient prescriptions [4, 7]. From a data privacy perspective, the RAG model poses two primary risks. The first risk involves data leakages that may occur when utilizing a third-party embedding service. The second risk is related to potential data privacy concerns that can arise when sending the relevant documents as prompt context to the LLM providers.

This paper aims to address the critical issue of data protection in RAG model by proposing a framework named RAG Local Placeholder Tagging - RLPT. The framework filters out sensitive information in context documents and user prompts before sending them to third-party embedding services or LLM service providers. The RLPT framework functions as a plug-in module for any RAG system, eliminating the need for modifications to the existing RAG system, embedding services, or LLM services.

The main component of RLPT is a private local LLM, which operates locally to identify sensitive information in user input and then replaces it with unique placeholders. These placeholders serve the purpose of indicating and concealing the actual sensitive data, preventing third-party service providers from capturing the original sensitive information during prompt processing. Once RLPT receives a response from these providers, it will substitute the placeholders with the original information before forwarding the response to the RAG system for further processing.

The RLPT framework addresses two main risks in the RAG system. Firstly, it addresses the risk associated with sending prompts to third-party LLMs by sanitizing the prompt. This involves filtering out any personally identifiable information before sending it to the third-party providers. Secondly, the framework helps mitigate the risk of data leakage that may occur when using a third-party embedding service. It achieves this by sanitizing the documents before they are chunked and sent to the embedding model. Additionally, the user query is redacted before it is embedded. This is important to prevent a mismatch between the entities in the query and the entities in the documents, which could result in the vector search failing to retrieve relevant documents.

The task of tagging sensitive information falls under Token Classification, also known as Named Entity Recognition (NER). To tackle this task, we utilize a pretrained LLM model to extract token features and employ Dense layers for token-level classification. In RLPT, we specifically utilize the DebertaV3 model for the local LLM. This large model consists of 24 layers and a hidden size of 1024. It has 304M backbone parameters and a vocabulary that contains 128K tokens. We fine-tune the model using a dataset of 22,000 synthesis essays, which includes seven types of PIIs such as name, email, username, ID number, phone number, personal URL, and street address.

We evaluate RLPT on a set of 4000 synthesized prompts. These prompts cover a wide range of information, including privacy-sensitive details like names, email, id number, street address, and web url. They also include non-privacy related prompts, such as generic questions and prompts without named entities. Our evaluation demonstrates that RLPT is highly effective in identifying PII in user prompts, achieving an accuracy rate of 88.7%.

Overall, the research contributions of this paper can be summarized as follows:

- We address the privacy issues in the operation of the RAG system.
- We have designed and implemented a framework called RLPT, which preserves user privacy when using RAG. RLPT is the first framework known to us that uses a LMM to address the two main privacy issues in the RAG system.
- We demonstrate the high efficiency and effectiveness of the RLPT framework and the LLM-based prompt sanitization mechanisms in protecting user privacy, with minimal impact on the user experience and without modifying the RAG system.

The rest of this paper is structured as follows, Sect. 2 introduces the privacy concerns and potential data leakages in the RAG model. Section 3 discusses recent published work in this field. Section 4 presents the proposed RLPT framework. Section 5 describes the implementation and evaluation of RLPT. Finally, Sect. 6 discusses limitations, future work, and concludes the paper.

2 Privacy Leakage in RAG

2.1 RAG Pipeline

As depicted in Fig. 1, the fundamental workflow for a RAG pipeline is outlined below:

- A user initiates an inquiry, referred to as the query.
- Based on the query, the relevant documents and paragraphs are retrieved from the knowledge base.
- The query and relevant documents are inputted into a LLM, with instructions to generate a response.
- The generated response is subsequently transmitted to the user. (Optionally, the relevant documents may also be shared with the user).

The construction of the knowledge base consists of the process of chunking and embedding the source data into vectors. This embedding procedure involves the utilization of an embedding model such as OpenAI's text embedding. Subsequently, these embeddings are stored in a vector store, wherein each embedding vector is associated with its corresponding chunk. The vector store facilitates the efficient identification of the most similar vectors when provided with a query vector.

In order to retrieve relevant documents based on a given query, the following steps are undertaken:

- Embedding the query into a vector.
- Determining the most similar vectors to our query vector.
- Retrieving the chunks that correspond to these identified vectors.

2.2 Potential Data Leakage

We assume that users are honest and are using a secure system and browser to access the RAG system. Additionally, the RAG system is willing to share user prompts and source data with third-party LLM service providers and embedding service providers. However, users have concerns about their personal privacy and do not want to burden the third-party providers with the responsibility of protecting their sensitive information in the prompts.

We consider third-party service providers, who offer LLM service and embedding service, reliable only for fulfilling user requests accurately. However, we do not trust them to withstand malicious attacks. With this assumption in mind, the RAG system communicates with users and third-party service providers through encrypted channels and utilizes industry-standard technologies to safeguard user data from unauthorized access. Nevertheless, third-party providers may collect and utilize user prompts and source data for business-related purposes, such as improvement and research. This may inadvertently lead to the disclosure of sensitive information through data breaches or model inversion attacks [8].

From a data privacy perspective, the RAG system has two main risks. Firstly, when constructing a RAG pipeline to effectively manage inquiries related to a company's internal documents and guides, it is necessary to chunk and embed these documents.

Embedding the documents often means forwarding the raw text into a third-party embedding provider like OpenAI or Cohere. This involves a huge privacy risk as it may expose confidential data such as dates, product names, specifications, source code, or project briefs. Furthermore, these documents may also contain personal information about the company's clients or employees, including names, dates of birth, social security numbers, or salary details. Additionally, the user query must also be transformed into embeddings as these embeddings will be used to retrieve the relevant sections. This introduces another potential risk to data privacy.

Secondly, the primary step in a RAG pipeline involves using an LLM to generate a response based on the query and relevant documents. In other words, when employing an LLM provider such as OpenAI or Anthropic, it is necessary to transmit the user query and all retrieved documents as a prompt to the LLM. This creates a heightened privacy risk within the RAG pipeline.

3 Related Work

3.1 Retrieval Augmented Generation

Retrieval-augmented generation, which was first introduced in [9] has become a popular approach for enhancing the generation ability of LLMs. RAG is a powerful tool that improves the accuracy and relevance of output while also reducing the occurrence of "hallucinations" in LLMs [2, 10]. Its flexible architecture allows for easy updates to the dataset, retriever, and LLM, all without the need for retraining. These advantages make RAG the preferred approach for applications such as personal chatbots and specialized domain experts [1].

However, the application of RAG also brings privacy issues. In [11] the authors have shown the privacy implications of retrieval-based LLM and identified privacy leakage of pretrained LLM [12]. The authors of [5, 13, 14] have shown that RAG is vulnerable to extraction attacks. The vulnerability of RAG makes its application in privacy domains under high risks.

3.2 Privacy Issues in LLM-Based System

As LLMs become increasingly integral to various applications, protecting sensitive data while maintaining model performance has become a priority in both academic and industrial settings. This section reviews key approaches such as Differential Privacy, Secure Multi-Party Computation, comprehensive privacy-preserving frameworks, and addresses the challenges in deploying these techniques in black-box LLMs.

Differential Privacy (DP) has emerged as one of the most widely explored approaches to safeguarding user data in machine learning models, including LLMs. DP provides mathematical guarantees that the inclusion or exclusion of a single data point does not significantly affect the model's output, thereby protecting individual privacy. Foundational works in [15, 16] applied DP to large-scale machine learning, demonstrating its potential for robust privacy protection. However, applying DP to LLMs involves significant trade-offs, particularly in terms of model utility. For instance, DP can reduce the accuracy or fluency of generated text, which is critical in many LLM applications [11]. To address these limitations, researchers have focused on refining DP techniques to optimize the balance between privacy and performance. This includes efforts to tailor noise addition mechanisms and develop adaptive privacy budgets, which can enhance utility while maintaining strong privacy guarantees.

Secure Multi-Party Computation (SMPC) is another promising technique adapted for use with LLMs. SMPC enables multiple parties to collaboratively compute a function over their inputs while keeping those inputs private. This makes it particularly valuable in scenarios where LLMs handle sensitive data across distributed environments. Seminal works in [17] laid the groundwork for integrating SMPC with machine learning, offering strong privacy guarantees. Despite its potential, the practical deployment of SMPC in LLMs faces challenges related to computational overhead and scalability. The increased complexity and resource demands of SMPC can limit its applicability in large-scale, real-world settings where LLMs need to operate efficiently. Moreover, SMPC's reliance on secure communication channels adds latency, which can be problematic for time-sensitive applications. As a result, while SMPC offers strong privacy, it often comes at the cost of reduced performance and scalability.

Given the limitations of individual privacy-preserving techniques, there has been growing interest in developing comprehensive frameworks that integrate multiple methods to enhance privacy protection in LLMs. Pioneering work in [18] on privacy-preserving deep learning set the stage for subsequent research, extending these principles to LLMs. Recent studies [19], have proposed hybrid frameworks combining DP, SMPC, and cryptographic techniques. These integrated frameworks aim to provide robust privacy protections while minimizing the impact on model performance. By leveraging the strengths of different approaches, such as combining the strong mathematical guarantees of DP with the collaborative privacy benefits of SMPC, these frameworks offer a more holistic approach to privacy-preserving LLMs. However, the complexity of these frameworks can introduce new challenges, such as increased computational requirements and the need for sophisticated implementation strategies.

A significant challenge in this area is the black-box nature of many commercial LLMs, where users have limited or no visibility into the model's internal workings. This opacity complicates the application of traditional privacy-preserving methods, which often rely on access to the model's architecture or parameters. Research in [20, 21] has highlighted the vulnerabilities of black-box models, including risks of model inversion and membership inference attacks that could expose sensitive information. The black-box nature of these models exacerbates privacy risks because it limits the ability to implement or verify the effectiveness of privacy-preserving techniques. Without transparency, users cannot fully assess how their data is being handled, which heightens the potential for data misuse or exposure. This challenge underscores the need for innovative approaches that can effectively protect privacy without requiring transparency into the model's internals.

While significant progress has been made in developing privacy-preserving techniques for LLMs, many challenges remain, particularly in optimizing the balance between data protection and model performance. Existing methods such as Differential

Privacy and Secure Multi-Party Computation provide valuable tools, but their application to LLMs, especially in black-box scenarios, often involves trade-offs.

4 Proposed Framework

4.1 System Overview

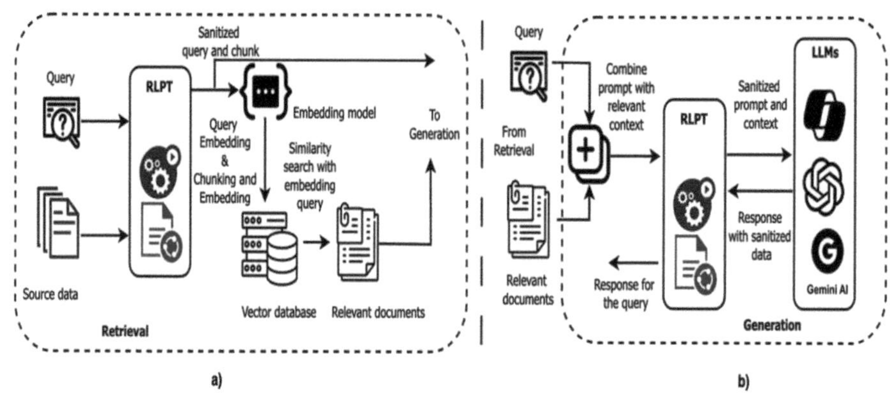

Fig. 2. System Overview of the RLPT Framework

The RLPT framework can help prevent sensitive data leakage in both the Retrieval module and the Generation module of a RAG system. As shown in Fig. 2a, the RLPT framework can be implemented in the Retrieval module to sanitize user queries and source data before sending it to a third-party embedding service provider. In this mode, all confidential information and sensitive data, such as PII, dates, product names, specifications, and source code, are redacted before being sent to the third-party embedding service. At the end of the Retrieval phase, both the user query and the relevant documents are sanitized and can be safely sent to the third-party LLM services provider during the Generation phase.

When the RAG system owner uses an embedding service deployed by themselves or a trusted party, it is not necessary to implement the RLPT framework in the Retrieval module. In Fig. 2b, the second implementation configuration, the RLPT framework is implemented in the Generation module to eliminate sensitive information in the prompt before sending it to third-party LLM service providers. It is worth noting that in this case, user queries themselves can contain sensitive information that requires redaction.

Table 1 shows the difference between the two implementation configurations. The details will be discussed in the next sections.

Table 1. Configuration for the implementation of the RLPT framework

	RLPT framework is implemented in	
	Retrieval module	Generation module
Operates on	The input to the embedding model	The input to the LLM
When to use	When using a third-party LLM API provider	When using a third-party embedding API provider
Batch/Online	This can be accomplished through offline batch processing, as it is possible for us to have the documents beforehand	This task can only be accomplished via online means due to the uncertainty surrounding the inputs prior to execution

4.2 RLPT Framework Architecture

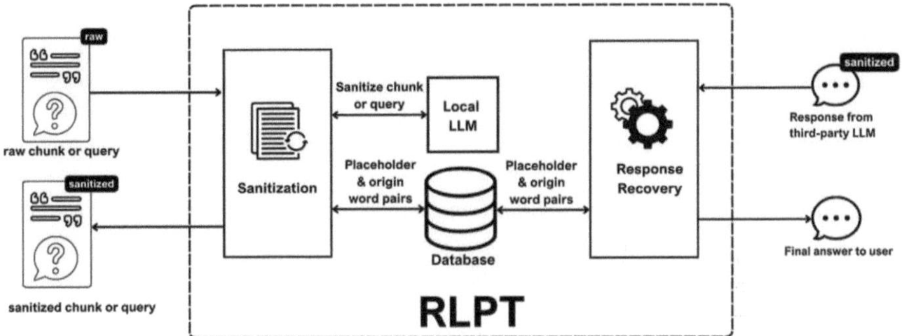

Fig. 3. The architecture of RLPT framework

Figure 3 depicts the main components on the RLPT framework. The input to the Sanitization component can be raw query from users, or raw chunk data source, or raw prompt created by the Retrieval module. The raw input are forward to the local LLM to detect and sanitize sensitive information within the text. The local LLM replaces sensitive entities with unique placeholders (e.g., <name_1>, <phone_1>), ensuring that the original data remains protected. This original data, along with its corresponding placeholders, is securely stored in a database for future retrieval. Once the sanitization process is complete, the sanitized query, documents, or prompt are transmitted to third-party embedding or LLM service providers for further processing.

The redacted responses from third-party LLM are processed by the Response Recovery component. It maps the placeholders in the responses with the corresponding items stored in the database. These mappings are used to restore the original data in place of the placeholders in the responses, creating the final answers. The final answers are then sent to the users.

4.3   Example of Raw Input and Sanitized Result

In the first implementation configuration (Fig. 2a), the raw input of RLPT can be a query from a user, for example, *"What is John Doe's phone number and email address?"* or a document chunk like, *"Mary Johnson owns a property in New York, and her contact number is 555-9876"*.

In the second implementation configuration (Fig. 2b), the RLPT's raw input is a prompt generated by the Retrieval module. A conventional prompt may resemble:

"Task: Answer the question using the provided context and question.

Context: Mary Johnson owns a property in New York, and her contact number is 555-9876

Question: What is Mary Jonhnson's phone number?"

In both implementation configurations, the RLPT framework will process the raw input and generate the redacted output. The output for the user query will be:

"What is <name_1>'s phone number and email address?".

The output for the document chunk will be:

"<name_1> owns a property in <location_1>, and her contact number is <phone_1>".

Lastly, the output for the prompt will be:

"Task: Answer the question using the provided context and question. The context and question contain placeholder tags to ensure privacy. Do not modify or replace the placeholder tags.

Context: <name_1> owns a property in <location_1>, and her contact number is <phone_1>

Question: What is <name_1> phone number?

Return: The answer based on the context and question.

Note: Placeholder tags must remain unchanged and should be used exactly as provided in your response."

When receiving the prompt above, the third-party LLM will response with sanitized text such as: *"<name_1> phone number is <phone_1>"*. This response will be processed by the Response Recovery component to get the final answer: *"Mary Jonhnson's phone number is 555-9876"*.

5   Implementation and Evaluation

5.1   Implementation

The core of RLTP is a local LLM. This section will discuss some key points regarding the implementation and training of the local LLM. In RLPT, we specifically utilize the transformer architecture and DebertaV3 model for the local LLM. This large model consists of 24 layers with a hidden size of 1024. It has 304M backbone parameters and a vocabulary that contains 128K tokens. Figure 4 illustrates the implementation of the local LLM:

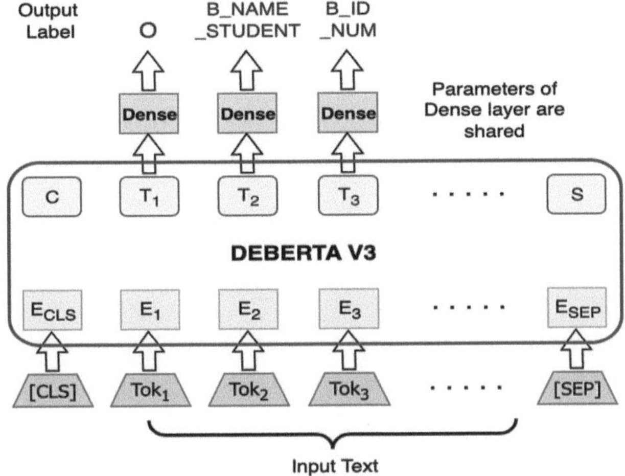

Fig. 4. Local LLM implementation

Dense layers have been added on top of the DebertaV3 model for token-level classification. The parameters of these Dense layers are shared. Unlike in text classification, the outputs of DebertaV3 are not pooled. Instead, the Dense layers are applied to the outputs in order to obtain label predictions. To clarify, the output of the DebertaV3 model is a 3D tensor with a shape of (`batch_size, seq_len, feat_dim`). The only change in this shape is in the `feat_dim`, while the other dimensions remain the same. After that, the Dense layer maps the `feat_dim` to `num_labels` and then applies a `softmax` activation to obtain the final prediction.

In order to fine-tune the LLM, we utilized a dataset obtained from a Kaggle competition[1]. This particular dataset consists of approximately 22,000 essays composed by students who were enrolled in a massively open online course. These essays were all written in response to a single assignment prompt, which required students to apply course material to address a real-world problem. Within the dataset, there are 7 types of PII, including: NAME_STUDENT, EMAIL, USERNAME, ID_NUM, PHONE_NUM, URL_PERSONAL, and STREET_ADDRESS. The labels for the PII are provided in the BIO (Beginning, Inner, Outer) format. To illustrate, consider the following sentence: "The email address of Michael Jordan is mjordan@nba.com."

In the `BIO` format, it would be annotated as shown in Table 2:

Table 2. Example of training data

No.	Word	Label	No	Word	Label
1	The	O	5	Michael	B-NAME_STUDENT
2	email	O	6	Jordan	I-NAME_STUDENT
3	address	O	7	is	O
4	of	O	8	mjordan@nba.com	B-EMAIL

---

[1] https://www.kaggle.com/competitions/pii-detection-removal-from-educational-data.

To optimize the model, the F_β score metric (with $\beta = 5$) is chosen. Additionally, the `CrossEntropy` loss, also known as log loss, was employed. This loss function can be defined as follows:

$$CrossEntropy = -\frac{1}{N} \sum_{i=1}^{N} \left[y_i \log \hat{y}_i + (1 - y_i) \log\left(1 - \hat{y}_i\right) \right] \tag{1}$$

where:

- N is the number of samples.
- y_i is the true label of the i_{th} sample.
- $\hat{y}_i$ is the predicted probability of the i_{th} sample being in the positive class.

The local LLM is implemented using TensorFlow version 2.15.0, Keras version 3.0.4, and KerasNLP version 0.7.0. For fine-tuning the model, we utilize two GPU T4 with 16GB RAM GDDR6. The source code for training and fine-tuning the model[2] can be downloaded from GitHub for testing purposes.

5.2 Evaluation

We evaluate RLPT on a set of 4000 synthesized prompts. These prompts cover a wide range of information, including privacy-sensitive details like names, email, id number, street address, and web url. They also include non-privacy related prompts, such as generic questions and prompts without named entities.

A set of 2000 prompts with privacy-sensitive data is collected from an external dataset in the Kaggle competition mentioned in Sect. 5.1. They are generated texts by OpenAI's GPT-4 Turbo model. An example query for GPT-4 is *"Arina Sun is a dental hygienist. Write a first person summary of something he solved in his job. Add the following information about him/her randomly inside the text: name is Arina Sun, email is arina-sun@gmail.net, address is 5701 North 67th Avenue."* The generated paragraph by GPT-4 for the example query is *"My name is Arina Sun, and I'm a dental hygienist. I have the privilege of assisting individuals in maintaining excellent oral health. One memorable incident that stands out in my career occurred during a routine checkup..."* This paragraph, which contains 210 words, is then stored as the data source text file for use in our evaluation.

To generate 2000 prompts without named entities, we utilize the following query: *"Generate 100 random questions (does not contain any personally identifiable information) to ask chatgpt"*. We choose to query for 100 prompts at a time to accommodate the limitations of the GPT-4 Turbo model. If the number of prompts is too large, the model may not generate them as requested. Afterward, we merge the two sets of prompts to create a set of 4000 synthesized prompts for evaluation purposes.

Table 3 shows the evaluation results:

---

[2] https://github.com/nxthang/RLPT-Framework.

Table 3. Evaluation results

	Predicted as Positive	Predicted as Negative
Actual Positive	92.7%	7.3%
Actual Negative	15.3%	84.7%

Our evaluation demonstrates that RLPT is highly effective in identifying PII in user prompts, achieving an accuracy rate of 88.7%. The false negative rate in the results (7.3%) mainly comes from the synthesized prompts that are poorly formed using the user names provided, such as *"hma"* or *"santang89"*. The false positive rate (15.3%) is higher than the false negative rate because some of the synthesized prompts without named entities use names of cities or countries in the prompts, such as *"What is the population of Tokyo?"*.

6   Conclusion

In this research, we present a new approach to address the critical issue of sensitive data protection in RAG model by proposing RLPT framework that filters out sensitive information in context documents and user prompts before sending them to third-party embedding services or LLM service providers. The evaluation results indicate that RLPT is capable of accurately detecting and filtering privacy and sensitive data, achieving a high accuracy rate of 88,7%.

The RLPT framework, while promising, still has a few limitations that need attention. First, generalization could be a challenge. Although RLPT has been tested on various types of sensitive data, its effectiveness in different industries or with varying privacy regulations is still uncertain. Next, a user's query may inadvertently reveal privacy-sensitive information, even if it does not contain any personally identifiable information or sensitive keywords. For instance, a user might pose medical-related questions about their health without explicitly using medical terminology. In these situations, RPLT may be unable to identify these privacy-sensitive topics.

To improve RLPT, future research could explore making it more scalable, developing more advanced models to handle unstructured data, and improving its ability to understand context. There's also room to adapt RLPT to allow users to customize their privacy preferences. These advancements could make RLPT a stronger, more adaptable solution for privacy in AI-driven systems.

References

1. Kamath, U., Keenan, K., Somers, G., Sorenson, S.: Retrieval-augmented generation. In: Kamath, U., Keenan, K., Somers, G., and Sorenson, S. (eds.) Large Language Models: A Deep Dive: Bridging Theory and Practice, pp. 275–313. Springer, Cham (2024). https://doi.org/10.1007/978-3-031-65647-7_7

2. Siriwardhana, S., Weerasekera, R., Wen, E., Kaluarachchi, T., Rana, R., Nanayakkara, S.: Improving the domain adaptation of retrieval augmented generation (RAG) models for open domain question answering. Trans. Assoc. Comput. Linguist. **11**, 1–17 (2023). https://doi.org/10.1162/tacl_a_00530

3. Dai, X., et al.: VistaRAG: toward safe and trustworthy autonomous driving through retrieval-augmented generation. IEEE Trans. Intell. Veh. **9**, 4579–4582 (2024). https://doi.org/10.1109/TIV.2024.3396450

4. Wang, C., Ong, J., Wang, C., Ong, H., Cheng, R., Ong, D.: Potential for GPT technology to optimize future clinical decision-making using retrieval-augmented generation. Ann. Biomed. Eng. **52**, 1115–1118 (2024). https://doi.org/10.1007/s10439-023-03327-6

5. Barnett, S., Kurniawan, S., Thudumu, S., Brannelly, Z., Abdelrazek, M.: Seven failure points when engineering a retrieval augmented generation system. In: Proceedings of the IEEE/ACM 3rd International Conference on AI Engineering - Software Engineering for AI, pp. 194–199. Association for Computing Machinery, New York (2024). https://doi.org/10.1145/3644815.3644945

6. Siddharth, L., Luo, J.: Retrieval augmented generation using engineering design knowledge. Knowl. Based Syst. **303**, 112410 (2024). https://doi.org/10.1016/j.knosys.2024.112410

7. Cyril, Z., et al.: Almanac — retrieval-augmented language models for clinical medicine. NEJM AI **1**, AIoa2300068 (2024). https://doi.org/10.1056/AIoa2300068

8. Fredrikson, M., Jha, S., Ristenpart, T.: Model inversion attacks that exploit confidence information and basic countermeasures. In: Proceedings of the 22nd ACM SIGSAC Conference on Computer and Communications Security, pp. 1322–1333. Association for Computing Machinery, New York (2015). https://doi.org/10.1145/2810103.2813677

9. Lewis, P., et al.: Retrieval-augmented generation for knowledge-intensive NLP tasks. In: Larochelle, H., Ranzato, M., Hadsell, R., Balcan, M.F., and Lin, H. (eds.) Advances in Neural Information Processing Systems, pp. 9459–9474. Curran Associates, Inc. (2020)

10. Izacard, G., Grave, E.: Leveraging passage retrieval with generative models for open domain question answering. In: Merlo, P., Tiedemann, J., Tsarfaty, R. (eds.) Proceedings of the 16th Conference of the European Chapter of the Association for Computational Linguistics: Main Volume, pp. 874–880. Association for Computational Linguistics, Online (2021). https://doi.org/10.18653/v1/2021.eacl-main.74

11. Yao, Y., Duan, J., Xu, K., Cai, Y., Sun, Z., Zhang, Y.: A survey on large language model (LLM) security and privacy: the good, the bad, and the ugly. High-Confidence Comput. **4**, 100211 (2024). https://doi.org/10.1016/j.hcc.2024.100211

12. Peris, C., et al.: Privacy in the time of language models. In: Proceedings of the Sixteenth ACM International Conference on Web Search and Data Mining, pp. 1291–1292. Association for Computing Machinery, New York (2023). https://doi.org/10.1145/3539597.3575792

13. Hu, H., Salcic, Z., Sun, L., Dobbie, G., Yu, P.S., Zhang, X.: Membership inference attacks on machine learning: a survey. ACM Comput. Surv. **54** (2022). https://doi.org/10.1145/3523273

14. Li, Z., Hong, J., Li, B., Wang, Z.: Shake to leak: fine-tuning diffusion models can amplify the generative privacy risk. In: 2024 IEEE Conference on Secure and Trustworthy Machine Learning (SaTML), pp. 18–32 (2024). https://doi.org/10.1109/SaTML59370.2024.00010

15. McMahan, B., Moore, E., Ramage, D., Hampson, S., Arcas, B.A.: Communication-efficient learning of deep networks from decentralized data. In: Singh, A., Zhu, J. (eds.) Proceedings of the 20th International Conference on Artificial Intelligence and Statistics, pp. 1273–1282. PMLR (2017)

16. Abadi, M., et al.: Deep learning with differential privacy. In: Proceedings of the 2016 ACM SIGSAC Conference on Computer and Communications Security, pp. 308–318. Association for Computing Machinery, New York (2016). https://doi.org/10.1145/2976749.2978318

17. Bonawitz, K., et al.: Practical secure aggregation for privacy-preserving machine learning. In: Proceedings of the 2017 ACM SIGSAC Conference on Computer and Communications Security, pp. 1175–1191. Association for Computing Machinery, New York (2017). https://doi.org/10.1145/3133956.3133982
18. Shokri, R., Shmatikov, V.: Privacy-preserving deep learning. In: Proceedings of the 22nd ACM SIGSAC Conference on Computer and Communications Security, pp. 1310–1321. Association for Computing Machinery, New York (2015). https://doi.org/10.1145/2810103.2813687
19. Zhang, X., et al.: PrivacyAsst: safeguarding user privacy in tool-using large language model agents. IEEE Trans. Dependable Secure Comput. 1–16 (2024). https://doi.org/10.1109/TDSC.2024.3372777
20. Zhou, B., et al.: DPDLLM: a black-box framework for detecting pre-training data from large language models. In: Ku, L.-W., Martins, A., Srikumar, V. (eds.) Findings of the Association for Computational Linguistics ACL 2024, pp. 644–653. Association for Computational Linguistics, Bangkok, Thailand and virtual meeting (2024)
21. Kim, S., Yun, S., Lee, H., Gubri, M., Yoon, S., Oh, S.J.: ProPILE: probing privacy leakage in large language models. In: Oh, A., Naumann, T., Globerson, A., Saenko, K., Hardt, M., Levine, S. (eds.) Advances in Neural Information Processing Systems, pp. 20750–20762. Curran Associates, Inc. (2023)

Author Index

S. Thomassey et al. (Eds.): RAIDS 2024, LNICST 673, pp. 247–248, 2026.
https://doi.org/10.1007/978-3-032-14055-5